Advance Praise for *Green Is Good:*

"If 'Greed is good' is a catchphrase for unbridled capitalism, then 'Green is good' should be its flipside: There are amazing economic opportunities for those who act with environmental responsibility. Brian F. Keane educates us that green is not only the color of a healthy planet, it is also the color of money."

—*Michael Douglas*

"Keane proves it's possible to still teach an old dog new tricks. This book is for those who know they can do more to go green but don't know where to start. A must-read for every generation."

—*Ed Asner*

"Green *is* good—and Keane is green!"

—*George Stephanopoulos*

"*Green Is Good* is a critical book for our age. Environmental degradation is the most important strategic challenge we face, but, unless people understand the issues and the opportunities at a personal level, we can't achieve the cultural shift necessary to create major change. Brian Keane expertly does this."

—*R. Seetharaman, CEO of Doha Bank*

"Credible, consumer-friendly . . . engaging writing . . . Keane expertly tells the clean energy story with humor and anecdotes likely to capture readers who want to learn more. . . . He takes us to a place where economic advantage and public good align. . . . *Green Is Good* offers an economic message that even Gekko might buy."

—*RealEnergyWriters.com*

"Brian Keane's practical, idea-filled book takes living 'green' from a good idea to a reality."

—*Douglas I. Foy, former Massachusetts secretary of Commonwealth Development and former president of the Conservation Law Foundation*

GREEN IS GOOD

Save Money, Make Money, and Help Your Community Profit from Clean Energy

BRIAN F. KEANE

LYONS PRESS
Guilford, Connecticut
An imprint of Globe Pequot Press

Lyons Press is an imprint of Globe Pequot Press.

Project editor: Meredith Dias
Text design: Libby Kingsbury
Layout: Maggie Peterson

Library of Congress Cataloging-in-Publication Data

Keane, Brian F.
Green is good : save money, make money, and help your community profit from clean energy / Brian F. Keane.
p. cm.
Includes index.
ISBN 978-0-7627-8068-6
1. Renewable energy sources. 2. Renewable energy sources—Economic aspects. 3. Clean energy industries. 4. Clean energy industries—Economic aspects. I. Title.
TJ808.K43 2013
333.79'4—dc23

2012018495

Printed in the United States of America

10 9 8 7 6 5 4 3 2 1

To Kate and our wonderful children—
Karenna, Jack, Tommy, and Julia

Green is good, but you guys are great!

The point is, ladies and gentlemen, that green, for lack of a better word, is good.

—With apologies to Gordon Gekko, *Wall Street*, 1987

CONTENTS

INTRODUCTION

It's the fall of 2002—a cold, damp day in a nondescript office park in a featureless suburban enclave just outside Hartford, Connecticut. The town is known officially as Rocky Hill, but it's little more than a Hartford bedroom community anchored by a Wal-Mart and peppered with chain restaurants. The office I'm here to visit houses the Connecticut Clean Energy Fund, a quasi-state agency created in 2000 to help bring the clean-energy industry to the Nutmeg State. Like other offices in the complex, it feels like a Motel 6—but looks can be deceiving. That's certainly the case here because behind this monotonous façade, the renewable energy industry is being born. From solar to wind to fuel cells, the Connecticut Clean Energy Fund is working aggressively to launch a revolution in how we power our lives.

I'm here to join the revolution, beginning with the unofficial launch party for SmartPower, the nonprofit I had started a few months earlier. Surrounding me are some fifty people, mostly men—as it was back then—representing a broad spectrum of wind, solar, and geothermal energy producers, suppliers, and brokers. It's a first-class, committed group, but most of them are waiting to see whether my fledgling organization is in their camp.

I start with Smartpower's birth parents: not only the Connecticut Clean Energy Fund, which is hosting the event, but also the Pew Charitable Trusts, the Rockefeller Brothers Fund, the Emily Hall Tremaine Foundation, the Surdna

Foundation, and the John Merck Fund. Why would some of the nation's largest and most forward-looking private foundations put money into our venture? Because we, and they, believe that clean-energy solutions require the same consumer marketing approaches as traditional brands such as McDonald's and Coca-Cola.

Americans say they want clean energy, I tell the group. They're well aware of renewable sources, but almost no one is using them. Why? Because we market clean energy the wrong way. But marketing can also lead the way out of the dilemma. Madison Avenue created a nation of obese children and multiple generations addicted to cigarettes. Why can't we apply those same advertising techniques to clean energy and energy efficiency?

It's the same argument SmartPower is still making, but on that fall day in Rocky Hill a decade ago I was making it for the first time before many of the biggest experts in the renewable-energy business—praying not to fall flat on my face.

Via PowerPoint, I explained what SmartPower could, did, and does do for these producers, suppliers, and brokers:

- Identify barriers that consumers face when considering a clean-energy or energy-efficiency purchase.

- Conduct consumer-market research that pinpoints the most effective messages and calls to action.

- Build partnerships with stakeholders, utility companies, state agencies, and the federal government to develop campaigns that connect consumers and prospective customers.

- Construct marketing tool kits, websites, and media campaigns that deliver clean-energy messages in a compelling, breakthrough way.

When I do these presentations today, I note that *Healthy Living* magazine—one of those freebies you pick up at the supermarket—recently called SmartPower "*Mad Men* for an eco-conscious generation." On that day, though, I surreptitiously checked my watch, clicked off the PowerPoint presentation, reminded my audience that they were selling products, not position papers, and called for questions and comments.

The first questioners lobbed softballs slowly over the center of the plate. Then a wind-power producer in a Pendleton work shirt, L.L. Bean duck shoes, and cargo pants (probably stuffed with granola bars) took the microphone and threw his best fastball.

"I enjoyed your presentation," he said deceptively. (Think of Dwight Schrute, in *The Office*.) "But are you saying that we're *not* supposed to talk about climate change or the environment when we're selling our wind power?"

"No," I said, "I'm saying that you don't *have* to. If you lead with climate change, it might hurt. We want to help you draw customers who are buying products for other reasons, such as—"

"Because I'm telling you," he interrupted, 'if someone doesn't believe that global warming is the number-one issue in this country, then I don't *want* to sell any of my energy to him!"

"If we really are going to combat global warming and climate change," I asked, "why wouldn't you want as many people as possible to buy your clean energy—for any reason they want?"

"Because I don't!" he announced.

We had reached, as they say, a standoff.

We're still there a decade later.

Energy so often is treated as an either-or issue. Either you are for wind, solar, or geothermal, or you are for fossil fuels or nuclear. Either you believe in global warming, or you reject it out of hand. Either you are a halo-sporting, Occupy progressive, or a wary Tea-Party conservative. But none of that is the case, or more accurately, it doesn't *have* to be.

Clean energy, green energy, whatever term you choose, needs a big tent if we're going to make progress for ourselves, our local, state, and national economies, or the environment. We have to make room under clean energy's spreading canvas dome for new technologies and old ones, for utility linemen and lost-in-the-clouds inventors, and for traditional power companies and start-ups creating prototypes that most of us can barely imagine. United we stand, as the old saying goes.

We even have to make room sometimes for what ticks us off the most. More than a few of my strident green friends, for example, see nothing but red when they encounter pro-coal billboards along the Pennsylvania Turnpike that read: "Wind dies. Sun sets. You need reliable, affordable, clean-coal electricity." One even adds: "Without coal, most cities would be dark." With one exception ("clean coal" is an oxymoron) those billboards are just stating the obvious. The sun that rises also sets. Gales become zephyrs, then become still. And, yes, as the grid exists now, most cities would be dark at least part of the day without fossil-fuel power.

Electricity itself is a good case in point of how all of this pulls together—or should. You can produce electricity in all kinds of ways, including shuffling your leather soles across a nylon rug and zapping your brother with a coat hanger or, emulating Ben Franklin, by flying a kite in a nor'easter. For large-scale commercial purposes, though, electricity derives from rotating magnets around a coil of copper wire. It's that simple and value-free. The magnets are agnostic as to what makes them rotate. Maybe it's steam—produced from the earth's natural heat or by burning coal—or petroleum or natural gas, or biomass or nuclear fission.

Or maybe what turns the rotors is wind, which unlike coal, petroleum, or nuclear is a primary action, not a secondary one, with no waste and minimal energy loss between action—air flows created by the spinning of the globe in combination with thermal highs and lows—and reaction: the TV comes on, the computer boots up, your smartphone gets charged. Solar panels operate differently. Nothing rotates. In most solar installations, photovoltaic cells do the bulk of the work, and current is produced directly, but the larger question is still the same: When we hit that still moment in the middle of the night, are fossil and fissile fuels our only hope?

The short answer is no. But keep in mind that this book is called *Green Is Good*, not *Everything Else Is Bad.* To be sure, the longer answer includes the need for our country to increase its energy portfolio. As we'll discuss later, we all use much more energy today than we did just fifteen years ago. As such, we desperately need more and more energy. Yet we still have partisans in Washington and across the country actively turning down the opportunity to create more.

But there's a more inclusive and more accurate answer to the still-of-the-night question: the national grid. When the sun sets in Philadelphia, it continues to shine for another three hours in Los Angeles. By the time Seattle wakes up, Atlanta has been bathing in sunlight for hours. When the wind stops blowing across the Dakotas, it's likely to start up across the Atlantic Coast. In between, woven through all of those moments, fossil and fissile fuels are heating water into steam that turns turbines that as a general rule assure the power will be on 24/7/365.

There are exceptions. Solar hot-water heaters sidestep the nighttime sweats because hot water doesn't have to be constantly heated. The heater, however it's powered, acts like a thermos holding in the heat. Geothermal also bypasses the no-light, no-wind boundaries. It's clean, constant, and potentially bountiful enough to power all of planet Earth. For now, though, unless we're going to retreat back before the age of steam or barrel forward with our environment in tatters, clean energy sources and fossil and fissile fuels, and their proponents on both sides, need to coexist and work together.

Solar, wind, and geothermal advocates need to respect the history and economic importance of coal, oil, natural gas, and nuclear. Those energy sources have boiled our water, lit our lamps, and kept our country running for a century and a half—and in coal's case long before that. They still form a critical part of the national economy. Similarly, coal and oil advocates need to understand that the energy needs of our nation are so huge that we need all hands on deck, stat. We need more wind, solar, hydro, and geothermal on the grid if for no other reason than that we keep using and needing more and more energy.

The sponsor of those nettlesome pro-coal billboards—Families Organized to Represent the Coal Industry—has some solid practical reasons for backing the ad campaign. The average Pennsylvania miner earns almost $65,000 a year, nearly 50 percent higher than the average wage for all private-sector industries in the state. That ain't chicken scratch.

But consider this: Pennsylvania has long been considered a coal state. We learn this every four years when presidential candidates travel there, walk with the miners, and listen to their stories. We also learn this every time a mine collapses, horribly reminding us how dangerous mining coal really is. As a coal state, Pennsylvania has about 15,000 jobs predicated on that industry. That's a good number of jobs—but it pales by comparison to the top private employer in the state: Wal-Mart. That's right, Wal-Mart is the top employer in Pennsylvania, with over 48,000 "associates," more than three times the number of mining-related jobs. Maybe we should stop calling Pennsylvania a coal state and start calling it a Wal-Mart state.

The point, though, is that while Pennsylvania has 15,000 coal jobs, over 5,000 Pennsylvanians make their livelihood in the wind industry. The Keystone State is famous for coal and rightly proud of its mining history. Yet measured by jobs, wind is already a third of the way there. Soon enough, Pennsylvania could be the Wind State.

The coal, oil, natural gas, and nuclear crowds need to recognize that all four power sources are ultimately an end-game. In many cases, extracting them becomes ever harder because it requires going ever deeper or transporting them across ever-greater expanses, or resorting to environmentally and geologically destabilizing techniques such as hydraulic fracturing—aka fracking—which is making headlines across the country. In the case of raw oil in particular, the geopolitical consequences become ever greater as well. In all four instances, the environmental cost will eventually prove unsustainable—maybe not this decade, maybe not this century, but our obligation as humans is to live more than in just the moment. We need to consider the world we want our grandchildren and their grandchildren to inhabit.

Traditional fossil fuels and their proponents are going to have to make peace with renewables and their supporters. In the most energy-savvy states, this is already happening. Take Texas.

No state looms larger in the national mythology when it comes to fossil fuels. From the movie classic *Giant* to the TV classic *Dallas,* we picture the Lone Star State with oil wells gushing out black gold. But as we'll see, no state is doing more to develop a wind industry, complete with infrastructure. Why? Because energy people understand the future of power sources, not just their present and past. They know that if the country is going to meet its energy needs, we need to consider every available resource, and they're placing their bets now on what the black gold of the future will be.

It's also past time to stop perpetuating the silly notion that wind, solar, and geothermal are somehow radically new, prohibitively rare, or totally unreliable. All of them are exceptionally old, amazingly abundant, and persistently dependable. Romans warmed themselves in geothermal baths; the great Greek engineer Heron constructed an organ powered by a windmill; and the sun kept the dinosaurs from freezing to death.

Every hour of every day, the sun provides the earth with as much energy as all of human civilization uses in an entire year.

Science deniers will never accept this fact nor the reality of greenhouse-gas accumulation and global climate change, just as militant eco-warriors would rather swallow their own tongues than acknowledge that fossil and fissile fuels and the people they employ remain a key part of the American economy *and* power grid. But if the rest of us in the middle can get beyond ideology, militancy, and ignorance, maybe *we* can have a rational discussion. In places, that discussion is already happening.

The granola guy in the Pendleton shirt and cargo pants—the one who grilled me during that launch party in Connecticut a decade ago? Today, he sells wind power to one of the biggest snack manufacturers in America. That bowl of chips might be full of trans-fats and dripping with nitrates, but whatever fuel was burned to produce it has been offset by whirling turbines rotating their way into clean energy way out on the Great Plains.

That's exactly why we began SmartPower: to bring reason to the energy debate, to move choices beyond ideology, to apply the Coca-Cola Test to everything from wind turbines to Compact Fluorescent Lamp (CFL) light bulbs. Does it taste good, feel good, look good? Will it do the job and add value to my life? Will I be cooler, smarter, or sexier if I buy this product? That's how the Coca-Cola Company sells Coke. That's what the real Mad Men on Madison Avenue get paid to think about. That's the world of value-based decisions and product competition in which clean energy has to operate.

Green *is* good—really good—but it's even better once we rationalize the topic, demystify the subject, banish politics from the discussion, and boil matters down to the everyday practicalities, vanities, aspirations, and illusions that compel us wherever we are. That's what I'm going to do in this book. I will demystify and depoliticize clean energy and green technology. I'll tell the stories of those who are succeeding in the renewable energy field and of those who have yet to realize the potential, and I'll give you the numbers and tools to help make sense of the background noise that accompanies your everyday consumer decisions.

Along the way, I'll also shine a light on the massive opportunity renewable energy represents and demonstrate ways that all of us—college students, apartment dwellers, homeowners, and small businesses—can seize the opportunity for ourselves. A renewable energy future is good for personal budgets, housing and rental bills, retirement portfolios, communities and states (in the form of lower costs and increased revenues), and ultimately our country. It doesn't hurt that it's also good for the environment.

For clean energy, the future is now, not tomorrow, not some point where everything converges on the horizon. America already produces enough clean energy to power every professional sports stadium in the country, or every hospital in the country, or every factory in the entire industrial Northeast, or every single home in twenty-two states. But there's so much more that can be done, and so much more to do. We need to seize the moment.

This book will help you do just that.

1 THE PET-ROCK CONUNDRUM

The Problem, Clean-Energy Solutions, The Missing Link, How to Break the Logjam

A great marketer, the saying goes, can sell air conditioners to Eskimos. A truly great one can sell air conditioners to Eskimos in the dead of winter. Then there's Gary Dahl, who exists in a league all his own. Dahl sold rocks to people who had backyards full of them.

For those too young to remember, back in April 1975, Dahl was sitting in a bar in Los Gatos, California, listening to friends drone on about their pets—their problems and perks, chewed-up shoes, vet visits, kennel bills, walking Rover in the rain, cleaning the litter box, fishing the dead guppies from the tank before the kids woke up—every mind-numbing detail. *What's with these people?* he kept wondering. Why would they put themselves through all this work and expense? Then it hit him: the perfect pet. Easier than a dog and without the slobber and late night walks, more predictable than a cat with its finicky moods and fussy eating. No nasty smells, no seasonal shedding, no expensive food or shots, no untimely demise. Thus was born the Pet Rock.

To Dahl, the idea at first was just a joke, but whether he had a vision or simply too much free time he took his bar gag and turned it into a huge success. Pet Rock sales peaked during Christmas 1975, but the fad endures still—and not only in Trivial Pursuit games. Original rocks fetch $15 and up on eBay. Not bad for a novelty item that sold for $3.95, accessories included, almost forty years ago.

Dahl's maybe-not-so-stupid idea took the country by storm just as Americans were clawing their way out from under Watergate, the end of the Vietnam War, and the aftermath of the tumultuous 1960s. But something else was going on as Dahl was inventing his Pet Rocks, something that looked forward, not backward, and dealt with an emerging problem—not one that had been staring the nation in the face for decades.

The oil embargo of 1972–73 instituted by the Organization of Petroleum Exporting Countries did more than spike gas prices sky high and make OPEC one of the scariest acronyms of modern times. Gasoline rationing and long lines snaking around service stations nationwide offered exactly the shot in the arm that alternative energy sources needed. The government shifted into high gear, working on ways to wean the country from its dangerous dependence on foreign oil. The best, most promising, and readily available of them all was residential solar power.

In January 1976, just as Gary Dahl was banking his first million, Jimmy Carter moved into the White House and soon after, solar panels sprouted on the roof of the Executive Mansion. Time, place, and circumstance had all come together. Solar had its name in lights for all the world to see, and it would never be quite so good again. Even today, almost four decades later, Americans have cumulatively bought more Pet Rocks than solar roof installations, solar water heaters, or any other combination of clean-energy home heating or cooling systems.

It's not that we have a dedicated bias against clean-energy sources, despite all the political posturing on the subject. If anything, just the opposite is true. In fact, SmartPower polling has been very consistent on this score. About 84 percent of us say we want to buy clean energy. That's six of every seven people, and all across the board: young and old, male and female, GED and PhD. When was the last time 84 percent of Americans agreed on *anything*? Not apple pie (in one poll just 53 percent of truckers said they like it), not George Washington (in a summer 2011 ElectionsMeter survey, 70 percent of respondents gave our first president credit for being a good politician; the rest didn't or weren't sure), and not even Mother Teresa. When Political Hotwire recently asked if Mother Teresa did a lot of good for people, only 71 percent said yes.

Nor is price the biggest stumbling block. Sure, for the cost of a solar water heater you can fill a room with carefree Pet Rocks—even at today's eBay prices. But as we'll see, prices have been falling sharply on clean-energy delivery systems even as prices for traditional fossil fuels have been rocketing through the roof.

But here's the challenge and the source of the conundrum that is the title of this chapter: While 84 percent of us say we want to buy solar, less than 3 percent of us actually have. More Pet Rocks than solar energy.

But why? Why would we buy something anyone can find for free in any vacant lot rather than an energy source we clearly want and believe we need?

There are three answers: Marketing. Marketing. Marketing.

But before we get there, let's take a more in-depth look at the issue itself.

Is There an Energy Problem?

The purpose of this book is not to convince you that the world is about to end or even that greenhouse-gas accumulation and global warming are rising precipitously. If calved glaciers the size of Connecticut haven't at least piqued your interest in alternative energy sources, odds are you're not reading this book in the first place. Besides, we don't need to go to the Arctic or Antarctica to find a significant problem. We keep using more and more power, and the power we use keeps getting more expensive. If nothing else, we need to source the energy that will satisfy our huge and growing power appetite.

New York University physicist Martin Hoffert and his co-authors spelled out the situation in a seminal 1998 article for *Nature* magazine. Hoffert readdressed the matter in a

2011 interview for PBS that ought to be required viewing. Not much had changed between the two presentations, and neither makes for cheery conversation.

By Hoffert's best calculations oil and natural-gas production will peak by the middle of this century. Both will essentially play out by century's end. That's less than ninety years away. That seems a long way off, but think of it another way: A child born today could see the end of fossil fuel as we know it. Coal will remain plentiful. Massive reserves—over 500 years' worth—lie just below the earth's surface, but coal burns less efficiently and releases more carbon dioxide into the atmosphere, which adds to greenhouse-gas emissions and will likely lead to a global environmental crisis (if you don't think we're in one already).

Other energy options include nuclear fission, biomass (organic matter used to fuel power plants), and, of course, clean energy such as wind, solar, and geothermal power. Nuclear—while not contributing to global warming—does pose other problems. Strip mining for uranium devastates the landscape and threatens water supplies. Light-water reactors, the most common nuclear design, remain vulnerable to natural and man-made disasters. If fuel rods accidentally become uncovered, they head into meltdown, releasing toxic levels of radiation into the environment. This is what happened at Chernobyl and Fukushima, and what the country barely avoided at Three Mile Island in Pennsylvania in 1979. Alternative reactor designs are safer, true, but according to Hoffert, Uranium 235, the fuel now used in nuclear reactors, while vastly more efficient and less energy intensive to process than oil and gas, will be largely exhausted by century's end as well. In short, we would be replacing a moderately efficient energy source that won't last with a more efficient one that won't last either and the dangers of which are harder to control.

Biomass sidesteps the renewability issue—it's either harvested or dead vegetation—and is almost, though not quite, carbon neutral. (The carbon released when it's converted to fuel roughly equals the carbon stored in it when it is living, growing matter.) Biomass also can be converted into power in multiple ways: burning, chemical reactions, and biochemically through natural fermentation. The problem with biomass isn't the material or the process, it's acreage. Today, all of humankind consumes roughly ten terawatts of constant power—that's 10 trillion watts. To do all that with, say, wheat would require planting 10 percent of the earth's land surface equal, in Hoffert's words, "to all the land that's used in human agriculture right now . . . The only things on the planet would be human beings and wheat."

How about other renewables, like wind and solar? Here the challenge isn't the danger of the materials or the space needed. The challenge is scale. If oil and natural gas run out by century's end, if nuclear isn't safely viable, if biomass requires too much arable land, wind and solar would have to take up the slack on a scale unimaginable to most of us today. We need to get moving on that now.

In their *Nature* article, Hoffert and his coauthors call for what amounts to an Apollo Project similar to what landed Americans on the moon. In this case, to paraphrase President

Kennedy, we choose to build clean energy not because it is easy . . . but because we need it.

So, is there an energy problem? You bet. No matter how you define the issue: as accumulated greenhouse-gas emissions, elevated global temperatures, rising sea levels, or simply raw availability. The power that recharges your laptop battery, heats your home, cooks your rump roast or Tofurky, chills your wine, or gets you to the party has to come from somewhere.

OK, There's an Energy Problem. Are There Clean-Energy Solutions?

This is an easy one: yes, plenty of them, in all sorts of shapes and sizes. Whether you're a gadget guy or gal who always has to have the latest thing, like President Obama, who had his own iPad 2 before it was released; a systems freak who sweats energy delivery more than the end-line products themselves; or a policy wonk who locks down on buy-in, green is very good these days.

Below is an Insiders' Guide tour of some of the products and players working to close the gap between our looming energy crisis and our abundant clean-energy resources:

- **Window treatments.** US e-Chromic, a Denver-based high-tech start-up, is developing an ultra-thin film that automatically dims windows as sunlight increases. By reflecting sunlight instead of letting it pass through to the interior, e-Chromic's film makes cooling an apartment, house, or office cheaper and less energy intensive, and you can turn the film itself on or off with the flip of a switch. The company estimates that there are 19.5 billion square feet of existing windows in US homes and commercial buildings—26.5 miles squared—and that its technology can reduce air-conditioning costs in summer months by 30 to 40 percent while simultaneously driving down greenhouse gas emissions. That's *big* savings. Maybe most heartening of all, the film uses electrochromic technology developed by the US Department of Energy's National Renewable Energy Laboratory. Beating up on the feds is an American sport, particularly during election cycles, but this outfit certainly earns its keep.

- **Self-regulating home grids.** Another Colorado-based outfit, Tendril, has gone from a small start-up to a global company in just a few years by creating products that put consumers in control of their own energy destinies. One product includes a small screen that lets you know how much energy you're consuming and when you're using it, and automated thermostats that coordinate house temperatures to take advantage of off-peak energy prices. These gadgets aren't designed simply for saving energy; they allow you to show off how smart you are. (Yes, they have an app for that!)

Tendril's recent affiliation with Whirlpool will allow appliances to do the same on their own. A Whirlpool refrigerator using a Tendril platform, for example, will be able to communicate directly with utilities, purchasing energy at the best rates and managing its own energy use to maximize the cheapest rates. Imagine what happens when not just individual homes but grade schools, universities, factories, and farms all do the same. Peak-energy demands will flatten as the appliances themselves determine what they need and how long to wait to secure the best rates. As peaks level off fewer new power plants need to come on line and fewer utilities lines need building. Rationalizing power usage has a snowball effect, and no human can be as rational as a machine.

• **Hybrid and electric cars.** This is, by now, old news. It's no longer the crazy guy down the street who is plugging in his car. The Toyota Prius hybrid primed the market. The Chevy Volt and Nissan Leaf, for example, have fulfilled the increasing demand for gas-electric and all-electric vehicles. What's new is the cultural shift of old icons—the VW microbus, which is being redesigned with an electric engine.

More than forty years ago, the VW Samba microbus led the way to the mud-slick fields near Woodstock. The still-in-development VW Bulli microbus—smaller than its predecessor but with the same distinctive boxy lines—could become a neo-hippie conveyance itself, but it's going to be leading the way to the greener fields of clean energy. Powered by two 20-kilowatt lithium batteries feeding a 114-horsepower electric engine, the Bulli has a projected 186-mile range, enough hustle to top 60 mph in under 12 seconds, and a top speed of nearly 90 mph. The old VW microbus was lucky to break 65 mph and guzzled gas to the tune of about 26 mpg. With the VW Bulli, you just plug in overnight while listening to crystal clear Dylan and Hendrix on your iPod.

Even more exciting is that *all* the major car manufacturers are building electric and hybrid cars, not just the ones you would have expected a decade ago to be out on the edge. HybridCars.com lists nearly one hundred models—hybrid and plug-in—either in or nearing production, ranging from the $104,000 super-luxury Lexus LS 600h L to the $13,300 Kia Soul.

• **More electric wheels.** The creator of the super-compact urban Smart Car is moving into electric bicycles in Europe—and with good reason. Electric bike sales there have grown by about 100,000 a year over the last half decade, to roughly 800,000 units annually, and the new ebike looks to be a prime contender in a booming market. Powered by a lithium-ion battery, the bike's 250-watt electric motor kicks in as a rider begins pedaling. After that, pedaling exertion depends on which of four power settings a rider opts for. At the minimal power setting and in generally flat conditions, the battery is good for almost two-and-a-half days before needing a recharge, and riders are never far from their creature comforts. Electric bikes are already showing up in New York City as well, ridden by food deliverymen. An integrated USB port—the "e" of ebike—fully supports smartphones, music, and entertainment systems, even GPS for rides away from home.

• **Solar heat.** Solar is probably the hottest clean-energy growth item of them all.

As I write, California-based MiaSolé is piggybacking on a low-interest government loan to build a 500,000-square-foot plant in Griffin, Georgia, soon to be the largest solar factory in the country. The facility will manufacture ultra-thin solar cells. Unlike traditional silicon solar cells, these films use almost no semi-conductor material—about 1 percent of what's used in silicon-based cells. That makes the film more efficient, less bulky, and easier to blend into an existing roofscape as well as considerably cheaper to produce, about half the price or less of its silicon cousins. The film's copper indium gallium selenide iteration (say that three times fast) has achieved in excess of 20 percent efficiency in laboratory testing. That is, it converts 20-plus percent of the sun's energy that it receives into electricity. Traditional solar panels have an average efficiency rate of about 15 percent.

Old-line multinationals are getting into the game as well. Dow Chemical, one of the largest corporations in the world, revved up its research and development engine a few years back, creating what *Time* magazine dubbed one of the best inventions of 2009: solar roof shingles so thin and flexible that they can be hammered to any roof by any roofer. Imagine what that will mean when new-home construction revives. As Dow CEO Andrew Liveris puts it, "There is a potential here for a $5 billion market. Keep in mind, this isn't just a market we're entering. It's one we're creating."

Creating, though, doesn't mean owning. The only constant of solar-energy products is that breakthrough technology is waiting around every corner. The Department of Energy, for example, has given a $150 million loan guarantee (yes, one of *those* loan guarantees) to 1336 Technologies, a Massachusetts company that has developed a technique for casting silicon cells to specific measurements instead of slicing them from a large block. The difference might sound minor, but it's not. Traditional solar-wafer manufacturing and slicing result in as much as a 50 percent loss of material. 1336 Technologies, by comparison, is a lean, clean, cell-to-spec casting machine.

Another new solar-tech company, Abound, has started making cells out of cadmium-telluride instead of silicon. Why? Because cadmium-telluride cells—despite sounding like a Colorado ski resort—are cheaper to make and generate more power than silicon cells. It's still unclear whether they're the clean-energy-paradigm buster that puts solar over the top, but they're a big step forward and, like all the other solar technologies mentioned above, they could help make solar panels truly cost-effective within this decade.

Collectively, these developments and many more promise to go a long way toward improving the value proposition of solar, and value is the real bottom line here. GPs (otherwise known as Gadget People) are going to buy self-regulating appliances and USB-enabled electric bikes because, well, they're gadget guys and gals who want bikes with apps. For most of us, though, buying clean energy will remain a distant dream—like watching the sun rise at Machu Picchu—so long as it's only about doing what's right for the environment. "Do I want this, or do I need it?" is the basic consumer question. For GPs *and* for the rest of us, gadget makers are starting to shift the needle from "want" to "need."

A clean-energy gizmo, though, is only as great as the clean energy that goes into it, which brings us to the next question:

How about Harvesting Power from Clean-Energy Sources?

Michael Skelly and his partners at Clean Line Energy, a Houston-based start-up, are erecting billions of dollars worth of electricity transmission lines between sparsely populated but highly windy parts of the country—the Oklahoma Panhandle, for instance—to big cities with a high demand for power. As with a lot of the clean-energy industry at this point, there's still a certain leap of faith involved in Skelly's work. The land to which Clean Line Energy is building doesn't yet have wind turbines on it, but without the transmission lines there's no way for a wind energy company to transport the energy to urban centers. It's the same bet that Kevin Costner's character makes in *Field of Dreams* when he erects his baseball stadium in an Iowa cornfield: "If you build it, they will come." But Skelly is no starry-eyed rookie. Before launching Clean Line, he built Horizon Wind Energy from the ground up, then sold the company for a $1 billion profit.

Within the decade, Skelly expects his transmission lines to translate into another big payday for him and his investors, and he has good reason for optimism. In 2008, in the midst of the dire global financial bust, the wind business boomed. That year alone, US wind power capacity grew by 50 percent, adding $17 billion to the economy and creating more than 85,000 jobs. It's barely slowed down since, even in the face of a mostly stagnant domestic and global economy. The Global Wind Energy Council expects installed wind capacity to exceed nuclear capacity by 2014. By 2020 global-installed wind capacity could reach 1,000 gigawatts.

A gigawatt is literally one billion watts. Think of a light bulb. Let's say your home is lit primarily with 75-watt bulbs. One gigawatt could power 13.3 million of them. Here's another way to look at it: The average person in the country needs about 1 kilowatt of constant energy for residential purposes—TVs, computers, refrigerator, hair dryer, etc. That means 1,000 gigawatts could power the home needs of one billion people. That doesn't include industrial and commercial usage—huge draws on the national grid—nor does it solve the problem of how to store wind energy so a homeowner or factory operator can draw the power down when needed. At least for now, that's the grid's job—balancing out resources and demand. Still, a renewable-power source for a billion people is no laughing matter.

Maybe because it lies beneath the earth's surface, geothermal energy gets less attention than wind power, but here, too, the potential is enormous and the product green, clean, and bountiful. Other parts of the world already know the secret of geothermal. The Philippines, El Salvador, and Iceland fill a quarter or more of their energy needs by tapping into the heat produced deep in the earth. The heat comes from the decay of radioactive materials such as uranium. Steam, created as water flows over these deep-earth hot spots, carries the heat to the surface. Yellowstone's famous geysers are perhaps the best-known American example of geothermal power at work.

On a percentage basis, we're not in the same league as Iceland or the others when it comes to capturing geothermal energy, but we do have more installed geothermal capacity than any other place on earth. In California alone, some forty geothermal plants are producing better than 5 percent of the state's energy needs—and private industry has barely begun to tap into total resources.

According to the US Geological Survey, conventional geothermal sources in the thirteen western states have the potential to produce 33,000 megawatts of power. But an emerging technology known as Enhanced Geothermal Systems (EGS) looks like a game changer. Traditionally, geothermal power has been harvested wherever it happened to create a natural reservoir, underground or at the surface. EGS, though, uses what's called hydroshearing to inject water at high pressure into the earth, opening passageways using natural fractures in the rock and creating artificial reservoirs that then fill with naturally hot water. In theory—yet to be tested fully in practice—this could mean that geothermal energy could be tapped virtually anywhere, a prospect that scientists think could swell total geothermal power to about 518 gigawatts, enough to meet the existing residential energy needs of the entire country.

Scientists at MIT have estimated that EGS technology could produce 10 percent of America's electricity within fifty years at prices competitive with fossil fuel and with little or none of the greenhouse-gas emissions that come with traditional energy production. Other credible estimates suggest that geothermal could supply us with clean-energy electricity for 30,000 years. That's not a typo: 30,000 years, and that's only in America. The Union of Concerned Scientists estimates that the heat contained within roughly the outermost five miles of the earth's surface equals 50,000 times more energy than all the globe's oil and natural gas resources combined.

Those are crazy, unproved numbers so far, sure, but is it any wonder that tech-savvy global corporate giants have started partnering with entrepreneurial start-up geothermal firms to help make this steam-dream a reality? In one such pairing, AltaRock Energy received a major cash infusion from Google—along with significant federal grants—to help with its EGS demonstration project at the Newberry National Volcanic Monument in central Oregon, about thirty miles south of Bend on the eastern slope of the Cascades. AltaRock executive Jim Turner calls geothermal a "pot of gold." Maybe, maybe not. Either way, AltaRock will be drilling down almost two miles, some 10,000 feet, to where natural radiation heats the earth to 500°F. With that kind of energy naturally available, do you really want to bet against the possibility?

You can find a far more humble example of the clean-energy imperative and how it's reshaping thinking in entrepreneurial ways on Randy Jordan's farm in Rutland, Massachusetts. In a classic example of turning a problem into a solution, Jordan and some fellow farmers are turning dirty cows into clean power.

The fact is, cattle can't help themselves. They take in the food available to them—silage, grasses, etc.—process some of it into milk if they're cows, and get rid of the rest by traditional evacuation methods, affectionately known among the six-and-under crowd in my household as "poop!" Their methane and manure aren't all that different from yours or mine, but while cows are only a quarter as numerous as humans (around 1.5 billion of them), they are *much* bigger. A newborn calf can weigh almost a hundred pounds while the heaviest recorded bull tipped the scales at nearly two tons. They also have much more of a pass-through diet, expelling about thirty gallons of waste per day. Put all that together, and it's easy to see why, according to the United Nations, livestock are the single biggest contributor to global climate change, "responsible for 18 percent of the greenhouse gases that cause global warming, more than cars, planes and all other forms of transport put together."

Farmer Jordan and the four other dairymen who with him make up an energy cooperative known as A Green Energy can't change their cows' digestive systems. But they are turning output to throughput with a wonderful device called the Digester, which occupies about one-and-a-half acres of Jordan's farm, along with two tanks, a pump, engine, and separator. Built at a cost of about $2 million—from a combination of loans and grants from the US Department of Agriculture (USDA) and the Massachusetts Technology Collaborative—the Digester and its supporting parts use the unlikely treasure trove of manure to produce electricity.

Initially, the Digester extracts methane gas from the manure, taking about six weeks from the time manure first enters a holding tank. The gas then powers a generator that produces electricity for the farm, with excess power sold back to the local utility. The remaining solid manure, minus methane, then becomes virtually odorless fertilizer.

Jordan's 300 mature cows yield about 10,000 gallons of manure daily. By running the Digester around the clock and adding waste from a local food manufacturer to the mix, he eventually expects to provide enough electricity to power 300 homes. That's one home per cow. Devote a million cows to the same cause, and you've powered the homes of a large city, and on the math goes—all before factoring in the vast environmental benefits of capturing and taming all that bovine methane. Win-win, in short, big time.

If cows, why not humans? NASA has been experimenting since the early 2000s with a microbe known as Geobacter that it believes can process human waste and turn it into electricity. The challenge that drove the space agency into the poop-to-energy business was what to do with all the waste generated by a six-man crew during a two-year trip to Mars—about six tons, most of it feces. Traditionally, astronaut waste has returned to earth with the astronauts, but six tons?

First discovered in the grim sludge of the Potomac River back in the mid-1980s, the Geobacter microbes not only break down waste materials, they also handily deliver electrons pulled from the waste material directly to fuel rods, which conduct them into a circuit, generating electricity. In theory, that process should also allow scientists to generate power while purifying wastewater.

If human waste, why not all waste itself? Waltham, Massachusetts–based 1st Energy is testing a shipping container–size device it calls the Green Energy Machine (GEM) that uses intense heat (600-plus°C) to convert dried, pelletized trash into synthetic gas that can then be modified for use in natural-gas engines or mixed with diesel to run generators. 1st Energy says that three tons of waste will keep a 100-kilowatt generator running for a full day, with a net output of 72 kilowatts after deducting the power to run the machine. Given that we Americans generate about 170 million tons of unrecycled municipal trash annually—half a ton per citizen—GEM and similar technologies have the potential to generate enormous amounts of energy while dramatically reducing the need for landfill.

There is a vast array of new clean-energy products and delivery systems either now available or in the pipeline. But continued innovation requires continued funding at multiple levels: governments, agencies, foundations, corporations, utilities, and so on. Transmission lines, geothermal demonstration projects, manure-to-electricity digesters and generators all require big up-front outlays before the first kilowatt flows through the line or the first BTU warms a house.

Which brings us to the next big question:

Are Big Spenders Really Climbing Aboard?

Here, too, the news is good. As we've seen corporate giants like Google and Dow Chemical are leaping into the fray with money, technical know-how, and long experience, and they're not alone. GE Wind is fighting European players such as Vestas, China's Sinovel, and India's Suzlon for market share in the wind-turbine sector.

Broaden "clean" to include cleaning up old energies, and the billionaire boys club grows accordingly. Bill Gates, Edgar Bronfman Jr., heir to the Seagram's fortune and head of

Warner Music, and Virgin Group founder and serial entrepreneur Richard Branson are all committing big dollars and large expectations to technology to capture the carbon dioxide produced by burning fossil fuels. The immediate goal is to reduce greenhouse-gas accumulation, and Branson, with Al Gore's assistance, is offering a $25 million prize to that end. But all three men are also hoping to find ways to recycle captured CO_2 into commercially viable products, including—near the extreme—a fuel of its own.

In the present, Ford's new all-electric Focus is elbowing its way into a still-evolving market that so far has mostly piggy-backed on Toyota's Prius platform. One effort to stand out from the crowd has the Focus teaming with solar provider SunPower with an option to install a 2.5-kilowatt solar system on the car's roof. You charge the car overnight like a regular electric vehicle. But during the day, when you need the air conditioner blasting or the heat working, the solar gives added energy to increase your mileage and efficiency. With an estimated add-on cost of $10,000 for the option, the Focus-SunPower partnership is only for very committed greens, at least for now. But this is what happens in healthy, free-market competition. Imitation begets innovation, and a good idea keeps getting better, cheaper, and more popular.

As we'll see in Chapter 4, the Pentagon has been taking the lead on many green initiatives. In fact, the Green Energy Machine described above is being field tested at Edwards Air Force Base in Southern California. The Department of Energy (DOE) has also become a major player, maybe *the* major player in green technology. For years, DOE's National

Renewable Energy Laboratory has been seeding university labs and the private sector with breakthrough discoveries such as the effect of clouds on solar power and cellular weaknesses that can be exploited to extract more easily the energy potential in nonedible biomass such as cornstalks and leaves. Another new DOE program lets start-up businesses license one of the more than 15,000 energy-related federal patents for only $1,000—an incredibly low-entry threshold that should draw thousands of entrepreneurs. Of all the DOE programs, though, none appears to have more promise than the oft-derided Advanced Research Projects Agency–Energy.

ARPA-E, as it's known, was modeled on the Defense Advanced Research Projects Agency (DARPA), the US Defense Department (DOD) incubator that launched everything from the Internet to F-117 Stealth technology. The energy clone, however, was slow to achieve liftoff. Created in 2007, it wasn't funded until $400 million flowed its way as part of the 2009 American Recovery and Reinvestment Act. It's been catching up in a hurry since then. Its March 2010 Energy Innovation Summit attracted some 1,700 participants, while its three rounds of funding to date have footed the bill for over a hundred on-the-edge explorations covering everything from electrofuels to thermodevices, words so new that most spellcheckers don't yet recognize them.

ARPA-E's scientific advancements and those of many other public and private sources are also starting to close the cost gap between traditional and renewable energy sources. A recent report from Ernst & Young predicts that the cost of solar energy will fall to $1 per watt by 2013, half of what a solar watt cost in 2009 and a third less than the average $1.50 per solar watt in 2011. *Bloomberg New Energy Finance* reported much the same in a June 2011 wrap-up of solar-energy hardware prices. At these trend lines, solar will soon be roughly on par with coal, a huge milestone shift in the economics of clean energy.

States and state-regulated utilities also have been taking an active and often leading role in cleaning up old energy sources and securing renewables for the future. Some examples:

In a multipronged attack, Arizona's largest utility company, Arizona Public Service, is shutting down the dirtiest units of one its largest coal-fired power plants, expanding the modern and more efficient parts of the plant, which will be equipped with additional emission controls—the first steps toward cleaner coal. APS is simultaneously investing big in solar power, creating the largest solar array in the world. That's enough power to support over 70,000 homes in Arizona. All told, the move will reduce the Arizona utility's emissions of carbon dioxide by 30 percent, nitrogen oxide emissions by 35 percent, mercury emissions by 61 percent, sulfur dioxide by 24 percent, and particulates by 45 percent. All the while taking advantage of a clean and renewable power source that the state has in overwhelming abundance: sunshine. Of more interest to the utility and its customers, they'll have persistent, green, reliable energy added straight to their grid.

Some will say, *Yeah, but it's a desert*. True, but why not use it?

Not to be outdone, Texas is pouring $5 billion into a transmission network to support the development of wind farms in West Texas and the Texas Panhandle. Construction is underway already on a number of lines, with dozens more in the pipeline, so to speak, awaiting approval. Opponents are plentiful, however. No one particularly wants a transmission line running through their viewshed. What's more, for the roughly twenty-five million Texans, the $5-billion price tag represents a capital outlay of roughly $200 per citizen—nothing for some, but too much for others. An independent study by the Perryman Group, however, found that a typical Texas residential power consumer will save between $160 and $355 annually in electricity bills once the lines have been built. The investment is also expected to create $30 billion in economic gains, $3.8 billion in additional gross product, 40,000 jobs, and $2 billion in additional revenue for state and local governments. Those are easy numbers to support.

Meanwhile, back in the Northeast, barely within sight on a clear day of some of the priciest vacation real estate in the free world, Cape Wind, America's first off-shore wind installation, continues to fight for its life, despite winning federal approval in April 2011.* The legal battle over the proposed 130-turbine, $2.62-billion wind farm—including nearly a dozen lawsuits seeking to halt its construction—has dominated the Cape Wind headlines, but the real story is what

* The latest setback as I write: A federal appeals court overturned the Federal Aviation Administration's ruling that Cape Wind's turbines posed no threat to local air traffic.

the facility could provide if it's ever allowed to be built: clean, renewable energy for some 200,000 homes in the power-starved Northeast while offsetting the consumption of 113 million gallons of heating oil annually. Nantucket was the nation's first energy community as the center of the production of whale oil, the sweet crude of centuries past. How appropriate if it could become a beacon of the new energy economy as well.

Sometimes we get so caught up in the process of transforming America's energy economy that we lose sight of what it would really mean to harness the power of renewable resources effectively. Here's a great reminder that factors in only one facet of the clean-energy arsenal: There are approximately 30 billion square feet of commercial rooftop space available in the US for solar panels. If we paneled those rooftops, we could generate approximately 150 gigawatts of electricity—enough to replace 100 coal-fired power plants.

Major big-box retailers like Wal-Mart and Kohl's have gotten this message. They're leading the way in converting their own roofs to solar power generators and, critically, locking in energy rates for years to come. They join a long line of entities—government, corporate, and nonprofit—getting behind clean energy because it's exciting, it's real, it's here, it's working, and it makes economic sense.

Yet there's still that pet-rock conundrum—the huge gap between those who support renewables in principle and those who support them in action.

What's the Hang-Up?

Some of my green friends lament what they see as a lack of sustained seriousness on the part of the federal government. Many individual departments and agencies merit gold stars for their efforts, as do many individual representatives and other political leaders, but a polity that can't even agree on the existence of man-made global warming—when virtually all serious research has proved that it is happening—simply can't present the united front or muster the sense of urgency that drives people to action.

It would be nice to have a consistent cheerleader in Washington and a consistent attitude toward clean energy. But we shouldn't just wait for the government to do that for us. Coca-Cola didn't need federal policies to concoct its secret recipe. McDonald's doesn't really care who the president is. People buy Happy Meals no matter who occupies 1600 Pennsylvania Avenue. That's the marketplace at work, and that's the playing field on which we need clean energy and energy efficiency to operate.

Another barrier often cited for the slow growth in renewable energy is the NIMBY problem—Not In My Back Yard. It takes many forms, but it's perhaps most prominent where wind turbines are concerned. Critics opine that they make too much noise, they're too ugly and unnatural, and they represent too great a danger to birds, bats, and other flying life forms.

Here, too, I agree in part. Even the threat of a NIMBY protest can scare off clean-energy entrepreneurs, and the squeakiest wheels always get media grease. But dig beneath the protesters, placards, and plaintiffs, and you'll find a much different story. Wind-turbine critics fall squarely in the minority even when you confine the survey sample to those most likely to be affected by turbines and towers.

Example: A poll of 1,200 urban and rural residents across Washington, Oregon, and Idaho, sponsored by the Northwest Health Foundation and several public radio stations, measured public opinion on wind energy being generated effectively in the respondents' own backyards. As one pollster summarized the results, "An overwhelming percentage—80 percent actually of residents of rural areas of the Northwest—support wind farms being developed within sight of their homes. What's more interesting is that 50 percent strongly—not just somewhat, but strongly—support this."

As frustrating as NIMBY issues and federal inconsistency can be, the clean-and-green's most ardent proponents are occasionally more frustrating still. You've seen these true believers, who prefer life in the 1700s rather than the 2010s; the ones who wear hemp, eat only organic, and chastise you for having air conditioning. These folks have helped cultivate the negative impression of clean energy and energy efficiency as something that only a certain type of person does. If you're not of that type, then why buy solar or be energy efficient?

This kind of holier-than-thou attitude is waning these days, but a kind of reverse-halo effect lingers. Too many people still see going green not solely in terms of product differentiation—do I want a solar-powered hot-water heater or an electric one?—but as a full-blown lifestyle choice. Today, the solar hot-water heater, tomorrow, free-range eggs and elderly tuna caught with hook and line only. Clean energy should be a values-based decision in the same way that paying extra for a pint of Ben & Jerry's or a new MacBook Pro is a values decision. People should want renewables, but they shouldn't have to feel the need to embrace an entire system that has nothing to do with how their water is heated or how their cell phones are charged.

Ask people how they feel about clean energy and you get the overwhelmingly positive results we've already seen. Change "clean" to "green," conjuring its association with a more strident environmentalism as well as echoes of Ralph Nader and his Green Party presidential campaigns of 1996 and 2000, and the numbers begin to fall. Morph "green" into "alternative," the preferred adjective of Republican strategist Karl Rove and most of the traditional energy sector, and you've just entered a parallel political universe where renewables are as out of the mainstream as pierced nipples. (After all, if something is alternative, then it is by definition out of the mainstream. Score one for Rove.)

Bring Al Gore into the equation, and you can quickly turn a sensible discussion about clean energy into an ideological flame war. That the former, much demonized and lauded vice president and Nobel laureate can still stir such passions speaks volumes about the need to turn down the rhetoric on renewables and turn up the common sense. (As we'll see later, another frequently demonized whipping boy, community organizing, happens to be absolutely the most effective way to lower total energy costs *and* provide clean power to broad swaths of people.)

Then there are the actual, practical problems that revolve around renewable energy:

- Many don't believe that it actually works. "What happens when the sun isn't shining or the wind isn't blowing?"

- Where to get it? You can't pick up a solar panel at Walgreens or the grocery store. In most states, you can't call your utility company and ask for wind power. Even when you want to buy green, odds are that you can't get it easily.

- Why bother when it's more expensive? This is a fascinating barrier because it breaks down in two ways: First, it is more expensive. Solar, wind, and so on are premium products, and right now they cost more money. But who's kidding whom here? We love premium products. The massive profit margins of Apple, Ben & Jerry's, Louis Vuitton, Starbucks, and plenty of other companies can attest to that. But the second part of the cost equation is the opportunity cost. People believe that if they buy clean energy then they also must buy an entire lifestyle.

- It's a lot easier to do nothing—especially when life is coming at you fast and hard. Because clean energy seems like too much work (see above), it gets crowded out by the surge of other daily work and purchases.

Indeed, the list of reasons that we don't use clean energy almost reads like a David Letterman "Top Ten List":

10. There's still more government talk than consistent action from state to state or even community to community.

9. NIMBY activists stir up more energy controversy than energy consensus.

8. Clean energy is still crawling out from under a years-long, holier-than-thou halo.

7. Opponents have politicized the subject, suppressed objective science, and demonized proponents.

6. People just don't have time for it.

5. People don't know *how* to buy clean energy.

4. They don't know *where* to buy it, either.

3. No one wants to be stuck with a clean-energy G4 when the G5 is coming out next week.

2. It costs too much—both monetarily and in opportunity cost. No one wants to expunge all synthetics and air conditioning from their lives merely to be considered worthy of owning solar or wind or geothermal.

1. People don't believe that clean energy actually works.

There you have it—the Dilemma of Clean Energy Dissected into Ten Easy Parts, right?

Wrong. One central element is still missing.

How to Break the Logjam

The products are fine. It's the presentation that's holding clean energy back.

Look at Gary Dahl. Once he decided to turn his barroom gag into a business, he didn't just run an ad for Pet Rocks and tell customers they could take 'em or leave 'em. No, it was all in the marketing. The pet arrived surrounded in straw in a carrier case with air holes in it so the rock could breathe. There was a leash, too, and a user manual. This is where Dahl really shone. "The Care and Training of Your Pet Rock" provided thirty-two pages of instructions on how properly to raise and care for it. Those lessons on teaching your pet rock to stay and sit are what sent the phenomenon viral—or what passed for viral in the mid-1970s. Newspapers picked up on the rocks, *Time* and *Newsweek* ran stories. A product with zero-inherent value whatsoever, other than kitsch, became a million-dollar hit.

Compare that to clean energy. Like rocks, wind, solar, and geothermal have been around forever, but that's where the comparison ends. Dahl made rocks *fun*. He gave people a reason to buy them, even if the reason was no more complicated than a friendly or ironic laugh. So far with renewables, instead of clever boxes and funny instruction manuals, we get pictures of starving polar bears and asthmatic kids—both serious topics, of course, but guilt and fear don't drive us consistently to the store.

The truth is, there are only two reasons that we buy anything—either we want it, or we need it. Take soap. You buy it because you need it—and in some cases because you want it. There in the soap aisle you find yourself facing tens of brands, some sold in double packs, others in colossal bricks, still others differentiated by content (oatmeal and honey) and packaging (modern, retro, outré). How do you choose? You make a value decision. You buy the soap that's on sale that week. Or you buy the soap that comes in a brick of eighteen. Or you buy the soap made with oatmeal and honey. Or you buy the soap you've been buying for the better part of your life. Each of those purchases is a value-based decision.

Some soaps market themselves as organic or sustainably produced. It might sound like the message is drifting into ideology, but remember: Every square inch of that package is advertising. The manufacturers are betting that you'll count sustainably made suds as yet another value point to spur your purchase. Maybe it is, maybe it isn't. But the essence of the transaction is simple: you needed soap; you had certain values in mind; you evaluated the soap based on those values; you bought soap.

So it needs to be with renewables in whatever form—as raw energy, solar films, smart-grid interfaces, and so on. We need to learn to see them not as a political issue, not as a wedge issue, not as a problem that requires legislation to resolve, but instead simply as a product, as something to buy, as something worth buying. We need to approach clean energy not like health-care reform but like soda, not like an Appropriations Bill, but like, well, pet rocks.

First, though, let's take a look at the cleanest, cheapest energy source of them all.

WHAT YOU CAN DO

BASIC:

- Understand that clean energy and traditional fuel sources aren't an either-or choice. They can and do exist side by side.
- If you don't already have one, buy an automated thermostat for your home.

INTERMEDIATE:

- Contact your utility and sign up for their clean energy option.
- Give your home an "Energy Assessment."
- Invest in a bike.

ADVANCED:

- Purchase only Energy Star appliances, which use significantly less energy than regular appliances.
- Replace key windows in your home with smart windows that have sun-blocking technology in them, which will help cut your cooling costs during hot weather.
- Purchase appliances that communicate directly with utility companies and can purchase energy at the best rates.
- The next time you're in the market for a vehicle, visit HybridCars.com and go hybrid—or, better yet, completely electric.

2 A TALE OF TWO HOUSES

False Efficiencies, Energy Deficiencies, Phantom Load, Turn Offs, and Turn Ons

What's the cleanest, most efficient kilowatt of energy going—the one with no greenhouse-gas emissions, free as the air you breathe? It's the kilowatt you *don't* use. If you had parents like mine, you probably saw that lesson in action every day.

It's Needham, Massachusetts, 1973, and the house I grew up in is *crowded*—two adults and eleven children born more or less on top of each other. I was the pivot child, number ten, the end of the "once-a-years" but not quite the end of the line. Even by Irish Catholic standards, we were prolific. A teacher once asked one of my brothers if our family lived in a shoe.

Our freestanding house had three stories; a finished basement, five bedrooms (most with bunk beds) and two-and-a-half baths. We took the idea of a Navy shower very seriously. You wet down. *Then* you turned the water off and lathered. *Then* you rinsed—in a hurry—because if you didn't you might freeze to death before you were done. Every spring, Dad and the oldest boys still living at home lifted off the heavy storm windows, hauled them to the basement, and brought up the lighter but still unwieldy full-window screens. Every fall, the reverse happened. Year by year, frames and window casements warped a little more and undid a fit that wasn't perfect to start. But the screens mostly kept the bugs out, and the storm windows mostly did the same with the cold. That was about as much as anyone dared expect.

Heat in our house came courtesy of a big oil-burning furnace in the basement and circulating hot-water baseboard radiators. Cooling came from the time-honored technique of opening windows. Somewhere in our neighborhood a huge attic-exhaust fan was probably sucking hot air from the top floor and drawing a cooling draft through lower windows, but that wasn't our house.

When the winters blew cold and blustery—remember, this is northeastern Massachusetts—Mom sometimes hung a blanket over a critical doorway to lessen the drafts. That many humans of various sizes, all walking around at 98.6°F, acted like a second heating unit. But if I polled my brothers and sisters, none of us would remember being too hot in winter, and of course the conglomerate heat effect worked against us in the summer, especially when the heat index matched body temperature, or worse.

True, we Keanes were a little over the top quantitatively, but qualitatively, at least where energy was concerned, we lived the way most Americans did back then. None of our appliances or other electrical appurtenances had an Energy

Star rating—TV in the basement; console AM-FM stereo in the living room with speakers the size of a mid-range brother or sister; electric stove with exposed burners that glowed red-hot; dishwasher and garbage disposal; freezer in the garage opened sparingly and never, upon pain of death, just to feel cold air on a suffocatingly hot afternoon; Mom's prized canister vacuum cleaner; a small radio always tuned to WHDH-AM to keep her company in the kitchen; and our electric jewel in the crown, the refrigerator, before which, at any given moment, stood at least three and sometimes as many as five insatiable teenagers desperately searching for anything to stuff into their ravenous mouths.

Energy efficient? Not on your life. Nor was the energy cheap. In 1973, when I turned six, America endured the first foreign-oil embargo and quickly got an energy czar in the person of William Simon. Rationing began. If the last digit of your license plate was an odd number, you could buy gas on odd-numbered days only, even numbers on even days. To cut down on gas, the nation slowed to a mandatory speed limit of 55 mph. Year-round daylight savings time took effect, too, for one year only, though, as it turned out since the sight of little kids trundling off to school in the dark proved too hard for parents to stomach. Nationwide, the average price of a gallon of gasoline rose 43 percent from May 1973 to June 1974. The raw figures seem miniscule now—38.5 cents per gallon to 55.1 cents—but keep in mind that real per-capita income then was about half what it is today and cars back then averaged just 13.3 mpg. As gas prices soared, so did heating oil. Across the ocean, in response to the global effects of the embargo, Britain's Prime Minister Ted Heath appealed to his countrymen to heat only a single room that winter, a measure of austerity that recalled the wartime deprivations of thirty years earlier (and probably an idea that our parents briefly considered).

It was, in short, real money and real pain. In a more rational world, we all might give thanks that we live today in far more energy-conscious, energy-savvy, energy-saving times. But that misses a larger point and brings us to the second house of this chapter's title.

My parents' house was built in 1964, not long before my family moved into it. My house now, brick with stone facing, dates from mid-twentieth century, one of tens of thousands of suburban tract houses built to accommodate the explosive growth of the nation's capital as America met the dual threat of Japan and Germany and rose to superpower status. My wife Kate and I do not engage in the semiannual storm-window and window-screen ritual of my parents. Like most of you, we slide the storm windows up in the spring and the screens down, on vinyl tracks, and reverse the process each fall. It's easy as can be and provides nearly an air lock when all the winter parts are in place.

Every one of our major appliances bears the Energy Star seal of approval. They are lighter, brighter, tighter, and smarter. If you could give a dishwasher, oven, or washer-dryer combo unit an IQ exam, our appliances would be off the chart. Our parents' comparables wouldn't even understand the test questions.

For heating and cooling, our parents had an oil-burning, beast-in-the-basement furnace and a simple thermostat on which they could set one temperature for day and one for night—although internal mechanics somehow prevented either setting from exceeding 70 degrees. My wife and I, by contrast, have two climate-control systems in a house no bigger than my childhood home but almost infinite control over what happens hour by hour, minute by minute, degree by degree in both climate systems.

We are also a smaller family—part of the national trend that has seen family size decrease from 3.42 members in 1975 to 3.16 in 2010. My wife and I have fewer children (four), though we maintained the Keane tradition of piling them one right after the other. Fewer bodies mean fewer eyeballs staring into the open refrigerator, fewer bedrooms to heat, less collective shower time (although the Navy shower rules are gone), and therefore less exertion from a hot-water heater already infinitely superior in every way to the one that heated the water of my childhood.

In another five years, the last of our kids will start school. My wife and I will be working our full-time jobs and the Keane house in Arlington, Virginia, will be, at least during school hours from September to June, what the Keane house in Needham, Massachusetts, almost never was: empty (not counting the dog). The house can rest then. It might even gloat a little over its sterling energy profile, the straight A's that its appliances are earning, those airtight seals and dual-climate systems, except it's all pretty much a fraud.

That drafty house of my childhood used much *less* energy than the one I currently occupy. That's right. The house where we hung blankets over the doorways used less energy than my smaller Energy Star house today. Unless you live in a prison cell, back-to-basics commune, or a yurt, the same is probably true of your own circumstances versus your parents'. Our houses represent so much of what is wrong—and easily correctible—about how we approach and consume energy today.

Generation Wars

Part of what made our parents better energy consumers was a steel will and tight fists. They survived the Great Depression—when you saved *everything*—but never forgot its lessons.

The son of Irish immigrants—his mother a domestic servant and his father a greenskeeper—my dad scampered up the ladder of success: World War II veteran, college on the GI Bill, and a career as a successful pharmaceutical salesman. But like so many other Americans of his era he never forgot how hard it once had been to earn a dime. When we left for church on Sunday mornings, the thirteen of us arrayed along the sidewalk like ducklings, not a single light was burning in our Needham house, nor a table fan whirring, nor radio playing. I can't swear to it, but in winter Dad probably lowered the thermostat a notch or two until we got back from Mass. It was only a couple of hours, but why waste heating oil?

Can I make the same claim? Can you? If you're roughly middle-aged, probably not. It's not that we are a wasteful generation, broadly interpreted, but we are in many ways a rebellious one. Our parents marched around the house flipping off lights as soon as a room was empty. They guarded the thermostat as if it contained the secret formula for eternal life. You're cold? Put on another sweater. You're hot? Lower the shades and block the sun. It worked, sort of. But for many Baby Boomers, it created a pervasive sense of deprivation. The place was always so gloomy!

That's not us. We're the bright-lights generation. Our kitchens glow like operating theaters. Cranked to maximum output, our sound systems shake the walls and can break up kidney stones. And winter? It might be miserably cold outside, but that's no reason that we can't dress at home as if sitting poolside at a tropical resort. Intellectually, we Gen Xers are far "greener" than our parents. Just compare our Priuses to their gas-guzzling Oldsmobile '88s. Unlike our parents, though, we have high expectations for our creature comforts, too. Sacrificing in the name of energy conservation is not built into our generational DNA.

This development would be more alarming if a shift wasn't swinging the pendulum back in the other direction, as happens from generation to generation to generation. Evidence of this shift is appearing all around us:

- In a 2008 UCLA survey of nearly 250,000 college freshman, 45 percent said that adopting green practices to protect the environment was essential or very important.

- A 2010 Princeton Review survey of more than 12,000 incoming college students and their parents similarly found that 65 percent said that knowing a college's commitment to the environment could influence their decision to apply to or attend the school.

- SmartPower has been encouraging this generational shift with a campaign called America's Greenest Campus. In 2009, the 470 participating colleges reduced their collective carbon footprint by more than nineteen million pounds, or roughly $4.25 million in energy-cost savings.

Just about everyone intuits these broad facts and, at some basic level, understands their importance. That's why 84 percent of Americans say they want clean-energy solutions. We children of the Greatest Generation might be careless about energy and natural resources. We might even set bad examples. But there's always the 3 percent side to keep in mind—the percentage of people who have already bought renewable—and successive generations often learn in spite of, rather than because of, their predecessors. With efficiency, clean energy, and renewables gaining momentum, that looks to be the case now—and *just* in the nick of time.

In 1950, the US population stood at just over 150 million and total residential energy consumption at about 6 trillion BTUs. By 2010, the population had a little more than doubled to just over 300 million, while residential energy consumption had grown almost four-fold to 22 trillion BTUs. Why such energy-use inflation? The more complete answer must take into account the fact that our houses, apartments, and dorm rooms also have so many more outlets and circuits for powering those appliances and for sucking energy off the grid.

Parkinson's Law

Born in 1909 in Durham, England, Cyril Northcote Parkinson was a noted naval historian, author of some sixty books, and winner—in 1935, via his doctoral dissertation—of the prestigious Julian Corbett Prize in Naval History. But it was his work on public administration and bureaucracies that cemented his reputation for future generations and, in particular, a single tongue-in-cheek law he formulated and later enshrined in a book: Work expands to fill the time available for its completion.

Endless variations exist on Parkinson's Law, and the law itself plays off Robert Boyle's famous seventeenth-century formulation about gases, which holds, loosely, that gas expands to fill the space available to it. Herewith my energy version of both principles:

Electronics expand to fill the outlets available for running or recharging them.

Mom had just three plug-in kitchen appliances: a coffee-maker, a toaster that worked serious overtime to feed eleven hungry mouths, and the little radio mentioned earlier. The toaster and coffeemaker sat on the kitchen counter, the radio next to the sink, tucked into an out-of-the-way corner.

All three of her appliances could probably fit inside my microwave, which is just the beginning of the story. My family and I have a toaster oven, food processor, blender, rice cooker, and vintage popcorn maker. Our one-cup coffeemaker even has its own appliance: a clock.

My parents couldn't have had all these things. Cuisinart and its food processors didn't come into existence until 1973 and didn't reach mass acceptance until the early 1980s. Anyone making rice at home in the early 1970s used a heavy pot on the stove top, and the same for popcorn. And who in her right mind would want just one cup of coffee when the percolator could make eight at a time? More important, though, our parents couldn't have plugged in any of our extra gadgets without first unplugging the items they used most.

My own family's kitchen has outlets all over the place and more multiplicity than that. Once upon a time, the average American kitchen had a single clock to tell the time—in my family's case, a battery-powered Westclox hanging from the wall. Today, the average American kitchen has four clocks—wall, oven, microwave, and coffeemaker—most of which never tell exactly the same time, so a given time is arrived at, Colbert-like, by a collective truthiness.

Once upon a time, too, an average kitchen had a single light source, usually a rectangular fluorescent ceiling fixture that came to full illumination only by strobing, maddening stages, if at all. We don't take those chances any more. Our kitchen, like so many today, glows like the Vegas Strip with eight recessed lights and more shining in from adjoining rooms. All these lights can be dimmed to control both illumination and power needs. In reality, they can be dimmed effectively only with the old wattage-gobbling incandescent bulbs. Variants of the newer energy-efficient compact fluorescent bulbs can be dimmed, but the light often appears harsh and wavers epileptically.

So when my wife says, "Turn on the kitchen light," I automatically turn on all eight of them. Alas, though, those eight lights only scratch the surface of our profligacy and the profligacy of most Americans.

My parents had that one big, boxy TV as yours probably did, too. My wife and I have two flat-screen televisions; one in our bedroom, another in the family room. Nationally, that puts us in the minority: 55 percent of American homes now have three TVs or more. (The average American home contains more TVs than people.) A TV per se, though, doesn't display the full picture of its energy usage because so many of them are no longer stand-alone appliances. Four decades ago we all had rabbit ears. Today we have cable converter boxes and satellite receivers that allow us to watch hundreds of channels that most of us have never seen and can't ever imagine watching. Most TVs have DVD players attached to them and, in addition, some also still even have VCRs attached (to watch all the old VHS tapes we accumulated). All of them plug into a power strip, without which long ago we would have run out of ways to draw down from the grid everything we need to keep our empire of distractions going.

Then there's the PlayStation or Wii, which alone has as many electronic components as existed in all of the 1970s.

Plus stationary computers and chargers for laptops, cell phones, e-readers, tablets, emergency flashlights, portable drills—all of them on power strips as well.

Australian electrical engineer Frank Bannigan invented the "electrical power-board" in 1972 but failed to secure a patent for what became the power strip, losing untold millions in royalties as a result. But if Bannigan hadn't done the job, the electrical or electronics industry would have invented it for him. Gadgets and power strips have a symbiotic relationship every bit as intense as that between a pro athlete and his agent, and even more profitable to both parties.

The multiplicity of outlets in modern houses and renovations allowed electrical gadgets to invade our lives and fill our spaces. Power strips expanded the outlets and thus the gadgetry exponentially. An average dormitory room today has more plug-in potential than the entire White House had back when Woodrow Wilson was commander-in-chief during World War I. If the national power grid had expanded as exponentially as the demands on it, we would all be sitting pretty today, but it didn't. It's nowhere near as expandable as the means for tapping into it. That's a real problem and a very compelling argument for efficiency. Or at the very least, a good reason to become energy smart.

10 Percent Off!

What would you do, what would you give up if you had to reduce your power usage by 10 percent?

The table below shows the average power consumption of common household appliances and other electronics. (Note that with some appliances, consumption varies depending on how fully in use they are. A stove with upper and lower ovens going and all the burners on might use 12,000 watts. The figure below assumes only partial use.) For those that apply to your life, multiply the wattage listed by average daily hours used. This will yield your average watt-hours per day for each device. Then add all the average watt-hours together to find your total watt-hours per day.

Total, in this instance, is approximate. These are select electronics, not everything that's plugged in, such as electric toothbrushes, for instance. Wattage also varies by type and brand—the figures above are mid-range in most cases—and devices are not generally 100 percent efficient. If you want to be really good about this, most calculations call for multiplying total watt-hours per day by 1.3 to arrive at a more realistic figure. But while it's a snapshot, the total above will give you a figure to work with.

Now the hard part.

Let's say your watt-hours-per-day total is 30,000—a moderate figure for a full house. Ten percent of that is 3,000 watt-hours, and that's what you need to cut. Where do you cut it? The fridge has to stay on. Are you washing and drying your clothes too frequently? Seems unlikely. How about the hair dryer? That's an energy-eating beast, but how few minutes a day can anyone use a hair blower without suffering terminal split ends? Maybe you could cool the house for an hour less each day. Or give ten 100-watt light bulbs three hours more rest. Or eat your casseroles colder, or wash the dishes by hand every other time. Or…

Appliance	Watts	x Hrs./Day	= Avg. Watt Hrs./Day
Refrig./freezer (20')	540	_______	_______
Coffee maker	800	_______	_______
Toaster	1,200	_______	_______
Blender	300	_______	_______
Microwave	1,000	_______	_______
Hot plate	1,200	_______	_______
Electric oven	4,000	_______	_______
Dishwasher	1,300	_______	_______
Electric water heater (40-gallon)	3,000	_______	_______
Washer	500	_______	_______
Dryer	4,000	_______	_______
Iron	1,000	_______	_______
Hair blower	1,000	_______	_______
Furnace blower	600	_______	_______
Portable heater	1,500	_______	_______
A/C (room)	1,000	_______	_______
A/C (central)	3,500	_______	_______
Computer (laptop)	35	_______	_______
Computer (desktop)	130	_______	_______
TV (25")	150	_______	_______
DVD	40	_______	_______
Satellite dish	30	_______	_______
Light bulbs	(listed wattage)	_______	_______
Total Watt-Hours Per Day			_______

The point is that energy diets are no easier than other forms of enforced deprivation. They entail choices, inconveniences, and sacrifices. Although in this case there's a far easier way to tithe on energy consumption: Stop using the energy that you aren't really using.

Phantom Loading

A recent Nielsen Ratings report tells us that, in the average American home, at least one television is on for six hours and forty-seven minutes a day.

That figure is wrong. The median television is on 24/7. That is, it never really turns off, and that applies to all two, three, or four or more TVs in *every* household.

This is yet another energy issue that our parents didn't have to face when we were young. The TV back then came on, as did the kitchen light, in stages: a little white circle at the center of the black screen, a spreading haze followed by a full-screen snowstorm, and finally, Howdy Doody or Mr. Rogers, right in our very own house!

How much time elapsed between turning on the TV and viable image and sound? Maybe thirty seconds. If we turned the TV on twice a day, that amounted to a lost minute daily, or maybe six hours annually of waiting for the idiot box to fire up—time actually not wasted since we often spent it talking in anticipation of what we were about to watch.

Today, all that sounds as quaint as chamber pots and home spittoons. Our TVs come on instantaneously, as do our computer screens, at least as a general rule. But here's a secret:

The reason isn't faster fire-up technology. The reason is that these appliances are *always* on, gobbling up energy even in their sleep, just waiting for us to notice them and summon them back to action.

In energy circles, this usage is called vampire power or phantom load. While we sleep or sit at our office desks or struggle through Econ 101, our appliances are sinking their teeth into our little corner of the national power grid, sucking it dry. Like a phantom, the phenomenon walks among us every day, invisible to the naked eye. Estimates indicate that as much as 10 percent of all household energy consumption disappears into phantom load.

Want to know what that means for you? Take your last electrical bill and move the decimal point one column to the left. That's your money that your appliances are bleeding from you, and you don't even know it.

How can that be, though? Take a look around. The evidence of phantom load is everywhere.

Our parents' phones were hard-wired and drew the little power they needed directly from the telephone line. You had to stay pretty much in one place to use them, but, when the electricity went out in a storm, it was mighty nice to be able to make a call. My landline and yours—if you still have one—can do things no one would have dreamt possible four decades ago: speed dial, caller ID, date and time stamps, voice mail, etc. What's more, we're free as birds to carry the handset into the bedroom or out into the yard while we talk.

But all that freedom comes at a cost.

We might use our landline only an hour or less a day, but those phones are on duty around the clock, around the calendar, tracking everything, and charging their batteries while they do so.

Consider the microwave. It was invented by accident in 1945, shortly before the end of the war, by Raytheon engineer Percy Spencer, who was testing a new vacuum tube dubbed a magnetron when he noticed that it had melted the candy bar in his pocket. Spencer exploded an egg next using the same microwave energy, then quick-cooked a hot dog—the "Speedy Weenie" project as it became known—and thus the history of the kitchen was changed forever, or maybe just the history of popcorn.

As you saw in the table a few pages back, microwaves are energy hogs, roughly comparable to a window-unit air conditioner. But what accounts for the greatest annual energy output of the average microwave? Not the popcorn, leftovers, or even reheated coffee. Hour for hour, day for day, we don't actually use microwaves all that much. The greatest annual energy expenditure of these miracles of engineering, these spawn of serendipitous discovery, goes to that little clock that occasionally doubles as a timer. It's consuming 6 watts 24/7. The microwave itself would have to run about nine minutes a day to top that.

And so it goes. So long as it's not shut down, your desktop computer is using, on average, 7.5 watts of electricity regardless of whether you're sitting in front of it. A laptop uses about half that. The modem is good for 3.5 watts; the wireless router, roughly the same; an inkjet printer, 7.6. If you have a desktop, a laptop, and the peripherals all going pretty much 24/7, it's equivalent to keeping a 25-watt bulb burning day and night.

Computers and modems can be shut down without undue hardship. Yes, you have to fire them back up, but you can scratch the dog's ears while you wait. Televisions are another matter, though. They and the cable boxes that feed them are almost impossible to shut down fully, short of pulling the plug, and for a good reason. In most instances, when you start the set back up, you have to cycle through all the channels so that the TV and cable box synch. It's even worse with satellite dishes where the receiver has to find the satellite first, and then cycle through.

When it comes to phantom loads, however, the televisions themselves have no equal. When your flat-screen is off but not unplugged, it's using as much power as that old, giant console TV your parents or grandparents used *when it was on.*

A cable converter box and a TV consume a bare minimum of 20 watts every minute of every day, around the clock. Throw in two more televisions—to reach the national mean of three per household—a DVD player; a DVR, which can't sleep if it's going to do its job; and at least one other peripheral that stands vigil day and night, and you're in the 50-watt range. Add that lightbulb to the 25-watt bulb burning from the computing equipment, and you're lighting a very bright room, a room that glows all night while you sleep, and all day while you're at work or school.

Then there are the small, sometimes miniscule phantom loads that collectively loom large:

- Those four kitchen clocks always counting: 24 watts.
- Four cordless phone stations: 10 watts.
- Two cell-phone chargers, even when no one is charging: 5 watts.
- One multipurpose battery charger, in use or not: 2 watts.
- Two radio alarm clocks: 5 watts.
- The electric toothbrush you use exactly five minutes every day if you are a dentist's dream patient: 1.6 watts.
- Three surge protectors (even protection from electricity isn't free): 1 watt.
- The doorbell used mostly by people selling makeup or distributing religious tracts: 5 watts.

Not every house, apartment, or dorm room has cell phones charging on top of laptops next to modems and routers. Plenty of people of our parents' generation still turn off lights and peripherals religiously. But the average American home invisibly leaks about 50 watts of electricity constantly, enough to light a room around the clock. Multiply that by the roughly 115 million US homes, and you're closing in on

5.7 million megawatts, roughly equal to the output of 3,600 midsize nuclear power plants.

Turn Offs

Happily, there are many, mostly pain-free ways to become a more efficient energy consumer.

The same power strips that turned our dorm rooms, apartments, and homes into giant electronic outlets with beds attached can also act as the first line of defense against power waste. Instead of shutting down four different phantom loads—the clocks on the microwave, coffeemaker, and kitchen MP3 player dock, say, plus the illuminated window on the toaster that tells you the current darkness setting—you can plug all four into a single power strip and push its off button when you leave for the day: 8 a.m. to 6 p.m, perhaps, and again from 9 p.m. to wake-up time. The appliances will last longer, and you won't have to pay for a few pounds of coal to keep your kitchen on red-alert readiness.

TVs and peripherals, as noted, are another matter. I'm betting that not even Al Gore turns off his flat-screens, cable boxes, DVD players, or DVR boxes every night before bedtime and then cheerfully waits for them all to cycle through the channels before tuning in to *Today* or *GMA*. But what if you're heading to the beach for two weeks or home for winter break? Do you turn the TVs off then—*really* off—so the vampires aren't slurping away on the national grid? If not, you might as well leave CNN blaring at top volume while you're gone. Hey, it might even mean saving money by canceling that expensive security system you have.

How about the game console? A Wii unplugged when you're gone is a little offering to the energy gods. One left off but still plugged in is like asking a neighborhood teen to stare into your refrigerator for an hour each evening while you're away. That analogy is especially apt when you consider that a generation ago, the refrigerator represented the biggest energy drawdown for an average American household. Today, that dubious honor goes to our televisions even though they're on only about a quarter of each day and actively watched far less than that.

The same logic applies to dorm rooms, home offices, and work offices, too. What is the modem sending, the wireless router routing, the desktop computer computing, the printer printing, or the fax faxing while you're skiing at Park City, hiking the Appalachian Trail, pampering yourself at a spa, or road-tripping with Springsteen on the iPod? Nothing you could possibly care about. Shut them down. Disable the secret energy thieves when you go away for any length of time. You can sort through the unread mail while you wait for *Modern Family* to boot up.

Refrigerators don't really qualify as phantom loads, especially since no one who already has an automatic ice maker is going to return voluntarily to the pre-automatic days when ice cubes required pouring water by hand into twistable plastic trays or those aluminum trays with handles you pulled up to loosen them, sort of. The water dispenser on the outside door probably does qualify as phantom, but refrigerators today are far more energy efficient than ever before. A contemporary Energy Star fridge uses half the energy of one built before 1993 and 40 percent less power than one built before 2001—important savings when refrigerators still account for a handy chunk of the average electricity bill.

But buying a new, hyperefficient refrigerator always raises the question of what to do with the older, less-efficient model you're replacing. Nationally, about one in five of us hauls it out to the garage or down to the basement, where—let's be honest—we keep it running full time to chill a couple of cases of beer or some white wine for that magic moment when three old college friends stop by to reminisce about the good ol' days, and you need a drink or two get through the memories. If you buy a new one but keep your old fridge, you're actually making the problem worse, not solving it.

Hot-water heaters aren't phantom loads either. There's no secret about what your hot-water heater is doing when you're not looking. But like other phantom-load appliances, this one expends the vast bulk of its energy blindly anticipating your hot water needs. In this case, that's a *lot* of energy—about 166 constant watts waiting for you to shower or to wash the dishes. Add that 166-watt bulb to the others in your phantom load room, and you're getting close to illuminating Versailles at ball time.

Solutions? On-demand hot water—common in Europe but less so here—heats H_2O in a hurry when you need it, but not at all in between. As you'll see later, solar hot-water heaters are also thoroughly efficient and one of the easiest ways to go green and clean. Plus, they make a big difference in the electric bill. Or you can go back to the future with what's known as "solar thermal," an entirely nonelectric solution in which the sun heats all household water. And, no, that doesn't mean living in Southern California, South Texas, or Florida. It's all a matter of design. Beijing lies roughly on the same parallel as Philadelphia, and yet the sun alone—and nothing else—heats 99 percent of all residential hot water in the Chinese capital. 99 percent!

Alternatively, you could attack the problem on the demand side by selling your teenagers. Seriously. The average active teenager—again, no kidding—spends just under forty minutes a day showering in the morning, after sports, and before bed. Had my siblings and I done that back in Needham, the shower would have run collectively for upward of four hours a day. Teens are constantly plugged in, constantly charging, constantly idling their cars and just about anything else, so it's paramount to bring high school and college students into the clean-energy fold, but for now, let's return to where the chapter began.

If the cleanest kilowatt of energy is the one never produced because you don't use it, then the dirtiest kilowatt of energy—whether generated by wind turbine, nuclear reactor, or coal-fired plant—is the one you buy and use without ever realizing it.

The economic argument for energy efficiency is clear. The sheer wastefulness of so much of our energy consumption is equally compelling, perhaps even more so, because this is something we really do have control over right now. An efficiency campaign that SmartPower helped launched in Bedford, New York, just north of New York City, has assisted twenty-one homeowners in lowering their energy use by 10 to 40 percent, saving them collectively $43,000 per year. Again, that's real money in the bank.

But energy efficiency is a matter of personal choice. You can't browbeat people into disabling their own phantom loads. Nor can you shame or legislate them into doing so. The green halo effect doesn't have that kind of power. But once you tap into a natural desire to save money and be smart about what you use and waste, an economic imperative takes hold, and that's where we go next: to some of the 3 percent of Americans who are acting on what 84 percent of us believe. They've got great stories to tell.

WHAT YOU CAN DO

BASIC:

- Plug your kitchen appliances into power strips and turn them off at night and when you're gone for the day.
- Calculate your household's total watt hours per day of electricity used to better understand your energy habits.

INTERMEDIATE:

- Link your computer(s) and peripherals to power strips and turn the strips off when you're not using the devices.
- Link your video entertainment components to a power strip that you switch off when you leave for the weekend or vacation.
- Ditch the booze fridge in the basement or garage.

ADVANCED:

- Replace your existing round-the-clock hot water heater with an on-demand or solar-powered unit.
- Sell your teenager(s).

3 GOING GREEN AT HOME

Renewable Solutions, Energy Education, and Solar Decathletes

In the last chapter, we saw two homes: the sprawling, leaky, energy-efficient one where a football team's worth of siblings and I came of age, and the air-tight, energy-deficient home where my wife, our four children, and I live now. In this chapter, we'll see a third house, although the word "house" hardly does the subject justice.

Brampton, one of the prides of Orange County, Virginia, according to its nomination for the National Register of Historic Places, sits on land acquired in the eighteenth century by Col. James Madison, father of the president, whose estate, Montpelier, lay nearby. In 1844, John Hancock Lee purchased nearly 400 acres from the Madison estate, later naming his new holding Buena Vista after the 1847 Mexican-American War battle, and began construction of a handsome Greek Revival residence that stands to this day.

As it turned out, J. H. Lee didn't stay long. The property changed hands several times in the 1850s until it came to rest with the Bryans, newspaper barons from Richmond, who renamed it Brampton after family estates in Wales. A few years later Confederate general J. E. B. Stuart, a Bryan family friend, established headquarters there, pitching his tent under a young tulip poplar that was known ever afterward as Stuart's Tree.

Stuart returned in 1863, set up headquarters at Brampton again, and took advantage of its high elevation to keep a close watch on the surrounding countryside. That winter, he was sound asleep with his wife when Union troops stormed the house, forcing the famous Southern general to scamper out a second-story window and down a nearby pine tree—more lore for an already historic house. Nearly eight decades later, the Bryan heirs sold Brampton and 262 acres to neighboring Woodberry Forest School, and another decade on, the school unloaded the house and roughly twelve immediate acres to Virginius Randolph Shackleford and his wife, my wife's great uncle and aunt. Their son, "Shack," Kate's cousin, lives at Brampton today.

Now, you might think that a guy who rides to the hounds and lives in an antebellum Greek Revival house might stick his nose up in the air, but cousin Shack isn't like that at all. He's a country lawyer and about as down to earth as a guy can be. His house is a beauty, to be sure, but it's also about 4,000 square feet inside and an absolute bear to heat, one reason that Kate, the kids, and I tend to drop in on Brampton only in the late spring or early fall. It's also brutal to air condition, which is why Shack historically hasn't tried to beat the heat other than with fans, shades, blinds, and other old-fashioned devices.

It was during one such visit two springs ago that Shack proudly marched me up to the attic and threw open the hatch to reveal a sea of blown-in insulation, the first thermal blanket ever to grace those ancient joists. "This place is going to be a lot easier to keep warm now," he proudly told me. *And cheaper*, I thought, with heating oil inching up to $4 per gallon.

We visited again in the spring of 2011 on a day that turned uncomfortably warm as we neared the house. *It's going to be sweltering inside*, I thought.

Instead, the place felt pleasantly comfortable.

"Central air?" I asked, surprised that Shack would treat himself to such a luxury.

"Nope," he said, "geothermal," as if he were discussing a new brand of ketchup he'd bought at the grocery store the day before.

You could have knocked me over with a solar cell. Shack rides to his office in a battered pickup that belches a cloud of blue smoke every time he fires it up. He would, I'm sure, much prefer bourbon to green tea or, perish the thought, açai.

"Why?" I asked.

"It just made sense," he said, and once he explained it, it certainly did.

Shack had been paying about $4,000 a year for heating oil just to keep the relatively few rooms he used in winter at a barely tolerable sixty degrees. Even that he supplemented with electric baseboard and oil-filled electric radiators. The entire tab probably came closer to $5,000 per winter, and that setup did nothing for him come summertime.

Shack also had plenty of what makes geothermal practical: land. But not just any land, land that's easy to work with a backhoe. A house Shack's size at his latitude (about the 38th parallel) might require a quarter mile of trenching about four to six feet deep. Into that trench goes plastic pipe, generally made of high density polyethylene because it's very durable and porous enough to allow heat to pass through its walls. Fluid—either water or some sort of antifreeze solution—then circulates through the pipe maze by a ground-source heat pump to take advantage of the natural ground temperature at that depth, about sixty degrees. In summer, the circulation carries heat and, indirectly, humidity from the house into the ground. In winter, the reverse takes place: the circulating fluid picks up the ground heat and carries it into the house, where a typical forced-air heating system distributes it.

Ground temperature obviously varies depending on whether you live in Death Valley or next to Glacier National Park in Montana, but the principle behind home geothermal is applicable anywhere. It's just a matter of taking advantage of what's already in the ground, waiting to be used. If you're not convinced, think of a cave. In the summer, the interior of a cave is almost always cooler than the air outside. and in the winter, caves are almost always warmer than outside. Like geothermal houses, caves are warmed and cooled by the ground that surrounds them.

Whether you're heating or cooling with geothermal, the system is basically free once installed, apart from the electricity to run the pump. In fact, studies show that about 70 percent of the energy expended in geothermal is renewable. As a bonus, you can also add to the system a "desuperheater," which uses the summer heat it draws out of the house to warm household water for free and, in winter, provides about half the energy for the same purpose.

But geothermal installation isn't cheap. The size of the space to be heated and the desired BTU output determine the amount of trenching required; the bigger the space, the more trenching you need. The heat pump, the key mechanical piece of the system, generally runs in excess of $10,000. In Shack's instance, the bill, soup-to-nuts, rang up to about $30,000. Virginia unfortunately offers no tax credits for this sort of improvement, and federal incentives had expired by the time he had the work done, so his payback time—the cost of the system divided by the annual fuel-oil cost, pre-geothermal—is roughly seven-and-a-half years, a little better than the national average on such systems. But Shack now gets summer cooling, which he never had before, for free.

Worth it? Shack has air conditioning in the summer where he had none, and he has a warm house in the winter. You tell me, who wouldn't value that? As for monetary value, generally speaking, the break-even point comes (as it did for Shack) about halfway through year seven. By then, you've recouped your investment in the system. After that, the benefits side of the equation goes into orbit. By year fifteen, you've doubled your investment—roughly equivalent to a 6.7 percent annual return—and by year thirty, if you get there, you've quadrupled your investment, a 13-plus percent annual return. That savings is "profit" on which you pay no tax.

Digging Deep and Going Collective

Not all geothermal stories are quite as dramatic as Brampton, but collectively they tell a remarkably consistent tale.

Jonathan and Linda Hamilton, for example, live in a Connecticut home without a name, but they have similar reasons to be excited. Faced with replacing their old oil furnace and air conditioning at an estimated cost of $20,000 for both units, the Hamiltons opted instead for a geothermal heat-pump system at a total cost of about $23,000; $9,000 for the trenching, $14,000 for the heat pump. But their new system also came with $3,000 in rebates, which made the effective cost difference zero. The savings, on the other hand, have been little short of amazing. Pre-geothermal, the Hamiltons were using about 700 gallons of heating oil annually, spending almost $3,000 when oil hits $4 a gallon. Not only has that cost vanished, but their electric usage "has gone down 30 percent," Linda Hamilton told a reporter. "We've seen a 20 percent decrease in our electric bill, and we have eighty gallons of hot water with our new system where we only had forty gallons last year." The Hamiltons are saving on the order of $4,000 a year, just about what cousin Shack is saving, with a similar payback calendar.

In Clarke County, Virginia, about seventy-five miles north of Brampton, near the top of the Shenandoah Valley, Janet Eltinge lacked both the land and the geology to do horizontal trenching for geothermal. The cottage that she was renovating and expanding sits on a limestone shelf, so Eltinge drilled down instead. So-called "vertical-loop" systems operate the same as horizontal-loop ones but require less piping because the farther down you go—and these systems typically go anywhere from 150 to 450 feet deep—the more stable the earth's temperature. The cost? A bit more than Shack's since drilling is more expensive than backhoe-trenching. Her savings and payback period stretched out more, too, but with almost as spectacular a long-range return by fifteen or twenty years down the road.

Other geothermal installations lay pipe in a pond or lake. Water stores and releases heat just as land does. Whether you go horizontal, vertical, or aquatic, geothermal is mostly a rural, exurban, or spacious suburban renewable solution. The same is true of wind. To be effective, wind turbines require an average minimal wind speed of about 7 to 9 mph. If your property meets that criterion, a small wind generator about the size of an old TV antenna can generate about 1 kilowatt of power—enough to keep maybe your electric car running, as we'll see in the next chapter.

Larger home systems—a hundred-foot tower, a blade twenty-five feet in diameter—might produce as much as 18,000 kilowatt-hours per year. Given that the average price of residential electricity in the country was twelve cents per kilowatt-hour in August 2011, that 18,000-kWh annual output has a market value of a little over $2,100. But the installation cost hovers in the $30,000 range, and, while you might be able to sell power back to your local utility when the wind is blowing, you'll need to purchase conventional power when the wind drops below 7 mph. Unless you're surrounded by plenty of breezy acreage, a full-size personal wind turbine is almost certainly impractical.

But pooling resources to take advantage of wind power is a viable idea whose time has already arrived.

Farm communities have an edge over the rest of the country when it comes to going green. People who grow up working the land naturally take advantage of whatever opportunity nature throws their way. Cattle, chickens, corn, sheep, sorghum, soybeans, sugar beets, wheat—if there's a market and conditions are right, a farmer will grow or breed it. If that crop happens to be wind, what's the big difference?

Apples are New York State's leading fruit crop. In fact, the Empire State ranks second nationally in apple production only behind Washington State. But apples are a hard business these days. The industry has been moving westward for at least two decades, first to Washington, then clear across the Pacific to China.

In New York, only about one in four farmers earns more than $100,000 a year, and many make significantly less. To keep in business, apple growers in upstate New York have turned their collective orchards into a two-tier farm. Apples still grow in the orchards at ground level and are harvested in season, but thirty feet above them wind energy is harvested and converted to electrical power year-round in giant wind turbines, a win-win situation that farmers across the state and around the nation are increasingly adopting. Thanks mostly to such two-tier arrangements, New York has been adding about 500 megawatts of new wind capacity annually for the last half decade along with the jobs that come with them.

Further out on the edge, micro-hydropower systems create electricity in the same way that large, river-spanning hydroelectric plants have been providing electricity ever since homes were first wired. Micro-hydroelectric generators resemble a low-power electric trolling motor, but this rotating propeller is creating energy instead of expending it. Instead of distributing the power via a grid, the generators charge batteries—typically 12, 24, or 48 volts—that store energy for home use. Micro-hydropower requires very specific conditions and often a number of permits as well. You need running water on your property or water that can be made to run. The more the water falls and the greater its volume, the more power you'll get. The formula goes as follows:

$$\frac{\text{head (feet) x flow (gpm)}}{10} = W$$

In plain English, the vertical distance that the water falls ("head," measured in feet) times the flow in gallons per minute, divided by 10, equals the system's output in watts. Which means 500 gallons per minute falling 10 feet would produce about 500 watts, enough to keep your average refrigerator running full time.

Clearly you need either a lot of water or a lot of fall—best of all, both—to satisfy all your energy needs with a home micro-hydropower system, but clean-energy sources are not mutually exclusive. Water can work in tandem with wind, solar, and geothermal.

Consider Edward Roe in Manhattan, Montana. Roe took advantage of the running water on his property to power two 24-volt submersible generators. He also installed two sets of 12,100-watt photovoltaic cells on the roof of his custom-built log home. He knew he wouldn't be able to get completely off the grid, but to narrow the gap he replaced all his old light bulbs with compact fluorescent ones, substituted a natural-gas dryer and range for his electric ones, and upgraded a thirty-year-old freezer with a new Energy Star model. Before these changes, the five-person Roe family was drawing down 33 kilowatt-hours of electricity each day from NorthWestern Energy, about equal to the national residential average. After the changes, the family now uses just 2.5 kWh from the grid. At twelve cents per kWh, that's a daily savings of $3.66—about how much you pay for that coffee drink at Starbucks—or over $1,300 annually.

Tom and Verona Chambers used a different recipe in the house they had built in Black River Falls, Wisconsin—solar panels on the roof and a geothermal ground-source heat-pump system—but achieved much the same end: no cost for gas or oil to heat the house and a much-reduced electric bill from the local utility.

When Matt and Kelly Grocoff set out to rehab their century-old house in Ann Arbor, Michigan, they really did go all out: geothermal heating, air conditioning, and water; sensors that automatically turn off lights in empty rooms; water-saving, dual-flush toilets (with half-flush and full-flush options) far more common in Europe than here; high-efficiency, 1.5-gallon-per-minute showerheads that Matt Grocoff estimates save them 11,000 gallons of hot water a year and more than $100 in annual energy and water costs; and so on down the line.

The total cost of restoration, not including their prodigious sweat equity, is about $48,000, but that includes everything—all the Energy Star appliances, state-of-the-art plumbing fixtures, and more. Before they renovated, their monthly winter gas bill alone averaged $350. The Grocoffs' total tab for heating, cooling, and hot water, as of 2010, was about $525 per year. They folded the cost of the geothermal and hot water systems into their mortgage, but when they subtract the savings generated by those systems, they have a positive cash flow of $60 a month.

Let the Sun Shine In

Wind turbines, micro-hydropower systems, geothermal installations, and the like are the rock stars of home renewables: splashy, out there, and sometimes unpredictable, given all the variables that go into the equation. Solar, by comparison, resembles Tony Bennett: older, more reliable, and almost certain to give a top-notch performance every time.

Solar works—that's a fact—and it's been working for a long time. A century ago, Philadelphia inventor Frank Shuman set up a parabolic sun collector, over 200 feet long and 13 feet wide, on the banks of the Nile, fitted it with a mechanical system for tracking the sun, and produced enough steam to pump 6,000 gallons of water a minute to irrigate the surrounding valley. World War I and other harsh realities dashed Shuman's dreams of filling 20,000 square miles of Egypt with his solar reflectors and turning the eastern Sahara into a garden spot, but in the hundred years since, his vast reflectors have shrunk into tiny photovoltaic cells and even reduced to film. Their properties have been thoroughly catalogued, their utility as renewable energy sources standardized. Show a seasoned solar contractor your house, let him study its orientation (preferably south-facing) and the pitch and square footage of the roof, and the contractor can tell you, pretty much to the watt, how much electrical power a given number of photovoltaic cells will yield.

Even here, though, there's plenty of flux these days. Stories like the Solyndra scandal that rocked the White House in the fall of 2011 grab headlines, but the real inside scoop is that solar is stronger than ever. As we saw in the first chapter, new technologies are constantly coming on line, including solar films, solar shingles, new stamping processes, and other ways of wringing ever-greater efficiency out of every solar cell. Add to that the growth of the industry and the sheer quantity of solar cells being manufactured, and what was long a seller's market has begun to shift in favor of buyers. Solyndra might have been mired in politics, but what sank the company was free-market forces. It couldn't compete with companies manufacturing solar cells better and cheaper. That's what we *want* to happen in a vibrant market. Factor in the rising cost of traditional fuels and utility-generated electricity, and the incentives to go solar trend upward even without state or federal subsidies.

Ed Boesel didn't need trenching, drilling, running water, or steady high winds to go green with his Massachusetts home, but his results are proof that the maxim about doing well by doing good pertains just as much when it comes to solar as it does to more exotic renewables. The solar power he recently installed on his home roof is saving him $100 a month on his energy bill, but he also expects to earn $3,000 per year by selling the excess power he's producing to Massachusetts utilities. Collectively that's an annual savings of $4,200. "The system . . . has done exactly what the company promised," Boesel says. "It's saving me serious money in energy costs, and I am actually *making* money!"

This is in New England, a region not exactly known for sun-drenched days. But regional variations in total sunshine matter less than you might expect. New England and the Rust Belt get about four hours and twenty minutes daily of what's called direct normal irradiance (DNI), the absolute best sunlight for converting to energy. By comparison, the allegedly sunny South gets about 4.5 hours of daily DNI. Translation: A modest 225-watt solar panel in Boston will produce about .945 kilowatt-hours of electricity on average, over the course of a year, with the bulk coming in the summer when days are longer. The same panel in Atlanta will yield 1.01 kWh on average, a not-so-whopping 7 percent more.

The Southwest leads the nation in direct normal irradiance with 5.5 hours a day, 30 percent more prime sunshine than in New England. (A tiny geographic pocket around Death Valley tops that with six hours of DNI, but they don't call it Death Valley for nothing.) The excessive heat that comes with high sunshine at low latitudes can compromise the full efficiency of solar panels, but in Arizona there are few, if any, complaints. In Pebble Creek, for instance, Ron and Diana Johnson were paying about $160 per month for electricity before they went solar in 2008; now, their bill averages about $20, despite higher rates. Just as satisfying, the solar system they use has been 10 percent *more* effective than advertised and maintenance free.

When SmartPower consultant Dru Bacon, also of Pebble Creek, had solar panels installed on his roof that same year, the cost per watt was $7.60. That is, for every kilowatt of energy generated by the solar installation, homeowners could expect to pay $7,600. Since then, prices have fallen to about $3 per watt, and rebates and other credits can drop that figure even more dramatically. For reasons we'll see in the next chapter, Bacon's utility, Arizona Public Service, gives a $1-per-watt rebate to anyone who goes solar; the state adds an additional $1,000 rebate; and, depending upon the mood of Congress, federal incentives have included up to a 30 percent federal tax credit on total installation costs. Throw all those numbers together, and the price of going solar in Arizona has fallen by over 50 percent in just the few years since Dru Bacon made the choice.

And that's only the beginning. In a November 2011 column in the *New York Times*, Nobel Prize–winning economist Paul Krugman suggested that the solar industry has reached such a critical mass of innovation and expansion that prices, adjusted for inflation, could begin to fall about 7 percent annually.

One of the other advantages to going solar is that you don't have to jump in all the way from day one. Once you start the trenching for geothermal, there's a strong incentive to keep going until you're installing a system that will meet all your energy needs. Micro-wind turbines can meet very specific and limited needs, but the big turbines don't come in graduated sizes. Solar offers plenty of easy way stations along the way, smallish steps that allow you to get comfortable with renewables before you make a big leap. Tiny solar panels can power outdoor lights for paths or accents. Freestanding panels can light an outbuilding, or small roof installations a particular room.

Solar hot-water heaters are the first time many renewable explorers push off from shore. The cost is reasonable, $5,000 to $7,000, and the payback is easily calculable. Water heaters are often a crisis purchase, but for the systems themselves, outside of a crisis, the entry fee is low, the savings are real, and the cool factor is high.

The distance between a solar hot-water heater and a full solar-roof installation seems infinitely more manageable than the distance between, say, a few clean outdoor lights and the big solar enchilada. A solar water heater moves you out of the 84 percent of people who say they want to go solar and into the 3 percent who actually have. It gives you a concrete foothold in the clean-energy game, which is invaluable.

Taking Possession

The greatest leap to make with clean energy and our own homes—whether historic landmarks, rural ranches, urban apartments, man caves, or dorm rooms—is the concept of ownership. We all know what we own: a car, sound system, laptop, baseball-card collection, or whatever it happens to be. Ownership conveys value, and because we own and know that we do, we tend to take care of the object in question: rotate the tires, get the warranty, buy insurance to protect the investment.

Energy, though, is a fuzzier concept. A solar array on the roof, geothermal trenching and piping, and an inverter in the closet that shows photovoltaic cells being fed into the household current—these are clear signs of energy ownership and in some ways the best argument for installing them. The power that drives your house becomes personal, sometimes deeply so.

An oil tank in the basement topped off at, say, 300 gallons acts as a constant and often painful reminder that you have invested heavily in the fuel that heats your house, but for most of us the ties that bind us to our wall outlets are weak. We act as though energy and power come with the house or apartment. It's there. You plug into it. The coffee brews, the washing machine spins, the toast browns. When you move on to your next domicile in our ever-migrating culture, you leave the power and energy behind and start over.

But that's not the case.

As long as you receive a utility bill, you are taking ownership of the power and energy you consume every time you flip a switch or light a burner. You make ownership decisions with every bulb you do or don't turn off when you leave the room, with every notch you turn the thermostat up or down—whether furnace or central air—or the hot-water temperature setting.

Whatever your energy source, your utility is selling power to you ounce by ounce, kilowatt by kilowatt, cubic foot by cubic foot. If you don't believe that's happening, stop paying the bill for a few months and watch how quickly you lose your ownership rights. No wonder so many of us have a love-hate-forget relationship with our utilities: We love them when the juice is flowing, hate them when the bills arrive, despise them when storms knock down the lines and blackouts reinforce our ultimate powerlessness, and forget about them the rest of the time.

Apartment dwellers—the majority of Americans, lest we forget—have a particularly hard time establishing any sense of ownership. After all, the only tangible point of connection to the power and energy they use is the bill itself. Even the wall plates and outlets technically belong to somebody else, often someone you've never met and never will.

Just about all utilities offer apartment dwellers and the rest of us who pay utility bills a backdoor into clean energy, a check box where for, say, $10 a month, we can request that our power come from renewables. (These are known as green tags or renewable energy certificates—RECs for short.) The clean power, of course, doesn't actually come straight into your apartment. It's added to the grid at the source and combined with all the other energy types that get put there. Still, this can provide a big boost to mass green-energy providers like wind farms.

A few utilities have even created a group incentive. If enough residents in a particular apartment building, complex, or neighborhood agree to a clean-energy surcharge, the utility will erect a small, solar-powered array somewhere nearby, so that subscribers can see the fruit of their sacrifice. In practice, this can be an exciting exercise for a local community group, church, or private business. But the bottom line is really this: People buy expensive things, including higher-than-average-cost energy, because they have a better perceived value. Creating a value for renewable energy and energy efficiency, beyond simply that it's good for the environment, is what will create a true clean-energy marketplace. Most people understand that equation as long as they feel they have an ownership stake in the matter, but that leaves out a large swath of the most profligate energy users of all.

Walking Up the Ladder

If it's hard to convince apartment dwellers to use energy efficiently—and it certainly can be—imagine how hard it is to make the same argument to those with absolutely no direct ownership stake at all in the power they consume: high-school students, college students in dorms, those who live in government housing, military families, and needy families, for example.

Military housing is a very interesting case these days. Generals and admirals have always lived well on base, but generations of lesser officers and particularly enlisted personnel have had to cope with barely acceptable and frequently substandard housing. That embarrassment began to end, though, in the late 1990s when the Pentagon acknowledged that it wasn't very good at housing and decided to privatize the whole business.

Today, just a few companies build virtually all-new military quarters and lease the homes back to the Pentagon, and by and large they do a very good job of it. Drive by a base with upgraded family housing, and you can easily see the difference. New on-base homes are aesthetically far more pleasing than before. They often feature a small deck of some kind. Yards offer some privacy and landscaping softens the rough spots. Inside, you're likely to find that symbol of middle-class elegance—granite countertops—as well as super-efficient Energy Star appliances. Yet one huge energy-consumption problem remains. As we've seen, the average American home has a phantom load of about 10 percent. In the new, privatized military houses, despite all their upgrades, that figure hovers around 17 percent, a staggering 70 percent higher.

No wonder the major military-base leaseholders want to create a campaign to bring phantom load for military families down to the national average. That might sound like a yawner, but serious money is on the table. Some of these leasing companies spend around $100 million annually on energy for their privatized military housing. Trimming that tab back by $7 million would rein in one heck of a cost monster.

So how can they accomplish this small miracle? The same way that SmartPower has successfully modified the energy behavior of thousands of high schoolers and dorm denizens: by educating and training, concentrating on what we can change, and ceding what we can't.

Case in point: computers and modems. Virtually all military families with a member serving overseas, especially in Afghanistan, leave their computers and modems on 24/7/365. Shutting them down when not in use would save literally millions of dollars, no sweat, but no one can or should propose doing that for one simple reason: the computers are on so that their loved ones overseas can Skype home at any time of day or night. Yes, those computers waiting for the Skype phone to ring are a massive national energy drain, but they are also a miracle of modern communications and a great humanitarian leap forward for military families.

Printers, however, are another matter. As we've seen, computers and printers are mostly joined at the hip. When one is on, the other almost certainly is, too. But why? How often do you need to print a document *at a moment's notice*? Or fax something, for that matter, if you even still have a fax machine? Without a little push, we just don't think of these simple fixes, or that our flat-screens are on 24/7 as well as our microwaves, coffee pots, sound systems, and Wiis. (Despite my kids' pleadings, no one ever needed to Wii on a moment's notice, either.)

Enlisted families tend to be young and child-rich. That means lots of dishes and plenty of clothes to wash. Happily, new privatized military housing comes with highly rated appliances, but an Energy Star dishwasher or washing machine in and of itself doesn't guarantee that it will be used efficiently. Users have to know to combine loads, try the non-heated drying setting for dishes, use the convection oven to shorten baking time, and so on. That's basic, mundane information but these moms and dads already have plenty of heavy-duty concerns on their minds.

The Pentagon is about to kick in some extra incentive for base-housed families to modify their energy behavior. Its Office of the Deputy Under Secretary of Defense Installations and Environment posts on its Military Housing Privatization website a list of answers to frequently asked questions, including one cautioning military families that eventually they will be asked to pay directly for utilities and further advising that "Housing occupants should also realize that they may need to become more active in operating their heating and cooling system to further increase efficiency and minimize costs." That is, take ownership of their energy usage. The Pentagon recently sweetened the deal with a project called SolarStrong, a billion-dollar program to install rooftop solar systems on up to 160,000 private military residences. In these tough times, there's no chance that the DOD would do that without an eye to big savings down the road.

But saving money—your money, my money, our/Uncle Sam's money—is a long way from the end of this story. Energy efficiency is important, vitally so, but it amounts just to the first step up the ladder that leads to energy renewables. If I can get you to turn off a light today, tomorrow you're more likely to exchange that incandescent bulb when it burns out for an energy-stingy compact florescent one. Having done that, you're also more likely to think that maybe I'm not talking smack about that flat-screen TV secretly gobbling up energy like a circuit-board vampire. Up the ladder you go from there until renewables seem possible, probable, and finally inevitable, whether we're talking a hot-water heater, a car, or home sweet home from top to bottom and stem to stern.

Topping Out

Washington, DC, can be a cynical place. The city constantly reminds you that politics is the art of compromise. Ideals are inspiring, but on implementation they tend to break down into grimy practical issues and frustrating baby steps. Grand schemes bottle up in committees or get held hostage as part of a political calculus that has little or nothing to do with the scheme itself. A friend told me not long ago about a talk she heard at the Smithsonian Institution's Wilson Center in the mid-1980s: a valedictory by a Reagan-era cold warrior retreating wearily back to Harvard. Bringing big ideas to Washington, he said, was like trying to have fun in town with a $10,000 bill. No one can make change.

Clean energy's foes are well funded and relentlessly on message. Renewables' friends, by contrast, often resemble a herd of well-meaning cats, meowing more or less in unison but impossible to corral. Our goal remains constant—to get people moving up the ladder I was just describing—but progress can seem glacial. Then every few years, I check the winning entries in the Department of Energy's not-quite-annual Solar Decathlon championship and realize that other people have been racing up not only the ladder

I was hoping they would climb, but also the next one and the next one.

The competition pits teams of college students from across the country and around the world against each other. International entrants in 2011 came from Belgium, Canada, China, and New Zealand. The goal is to generate a building design that uses renewables in an attractive, affordable, practical, and cutting-edge way. Teams of students spend months perfecting their plans, then gather in West Potomac Park, just south of the National Mall, where they construct their model homes for all to see.

These are not run-of-the-mill college kids. Instead of partying in Cancun or supplementing their spending money by working as part-time baristas, solar decathletes devote their spare hours, including school holidays, to imagining and executing the homes of the future. Many don't receive college credit for their labors, but regardless of whether they're boosting their GPAs these students are designing some of the neatest houses imaginable, powered by all manner of solar renewables and equipped with some of the fanciest efficiency gadgets I've ever seen.

The Solar Decathlon began the same year as SmartPower, and it's been a blow-away experience at every turn. But the 2011 exhibition launched the competition to a whole new level. The technology wasn't just new, it was completely original and a paradigm jump. The same held true with the designs. Some of the model homes made Philip Johnson's famous Glass House in Connecticut look dull by comparison. Nothing was for sale. Everything was in the name of science, technology, and the limitless possibilities of renewables and imagination. But walking among the projects I still had to fight the powerful urge to cut a check for a down payment.

From the University of Maryland's liquid desiccant waterfall, a humidity-control system that takes the place of traditional air conditioning and doubles as a dazzling design element, to Middlebury College's in-kitchen garden (year-round growing, even in chilly Vermont) and the University of Illinois at Champaign-Urbana's solar awning (shade on the south side of the house and power from the shade), many design elements were completely new to me.

Purdue University created a "biowall," in which air for the HVAC filtered through an all-plant wall. The University of Tennessee designed color-changing LED mood lighting. Students from Victoria University in Wellington, New Zealand, constructed a drying cupboard, in which solar-heated water pumped through a heat exchanger does the job of a traditional, power-eating dryer. With their poplar-bark siding, Appalachian State University was among a number of teams that used phase-change materials to collect heat during the day and release it during the night. A form of paraffin also served that purpose for several competitors—wax that takes heat in, preserves it, and releases it on a predictable schedule. The Ohio State team went a step further with a complex system that tracked US Weather Service data, adjusted itself accordingly, and could store heat for up to thirty hours.

Charmingly, each of the teams reflected the particular microculture it represents. The City College of New York's Solar Roofpod, for instance, was designed to fit on the top of a standard New York City mid-rise building—a place where green-minded urbanites can produce their own solar power, plant a rooftop garden, and retain and recycle storm water. As a bonus, the roofpod also distributes power to the rest of the building. In the words of the students themselves,

> The Solar Roofpod is a "penthouse with a purpose," designed to respond to the market for economical new housing in cities. The target market is urbane, ethnically diverse, and progressive singles, couples, and young families. Other urban market segments that could find the Solar Roofpod lifestyle and its considerable utilities savings appealing are empty-nester cosmopolitans and immigrant families.

The New Zealand entry used recycled sheep's wool as insulation. China's entry, from Shanghai's highly regarded Tongji University, was a solar confabulation constructed from six recycled shipping containers, reflecting China's dominance in world trade. The joint entry of the Southern California Institute of Architecture and the California Institute of Technology resembled a movie set for the sequel to *2001: A Space Odyssey*. SCI-Arc/CalTech, as the team is known, defied convention by wrapping insulation on the *outside* of its entry. Flexible vinyl wrap does a great job of sealing the "CHIP" house—Compact Hyper-Insulated Prototype—but it looked a little like squaring off the Michelin Man. Attractive? Maybe not, but daring.

Is any of this headed for your neighborhood soon? Doubtful, although many of the prototypes are being reconstructed on teams' home turf. But the point is that every one of the entrants pushed the envelope of expectations and walked us collectively farther up the ladder of what efficiency, clean energy, and unfettered imagination can accomplish. Maybe only our grandchildren will have a chance to live in these green houses of tomorrow, but knowing they're there, waiting down the line, offers comfort. Fossil fuel has nothing but bluster to match this kind of innovation.

Then again, maybe the next time that Kate and I drop in on Brampton, cousin Shack will be posing in front of his biowall or touting his new paraffin-based phase change unit or zero-power drying cupboard. Once you go geothermal, just about anything is possible.

WHAT YOU CAN DO

BASIC:

- Investigate whether geothermal makes sense for you.
- Dip your toe into the world of solar power with tiny panels for outdoor path or accent lights.

INTERMEDIATE:

- Using the Department of Energy's website, determine if you can take advantage of any local, state, or federal subsidies or rebates to defray the cost of going renewable.
- Check your utility bill, visit the utility's website, or contact it directly to find out if it has a clean-energy surcharge or incentive program. If it does, sign up.

ADVANCED:

- Call a local solar contractor for an estimate, based on the position, pitch, and square footage of your roof, on how much electrical power you could generate and how much it would cost to install a system.
- When your old hot-water heater breaks—and it will—invest in a solar hot-water heater.

4 TAKING CLEAN ENERGY NATIONAL

Community Action, Turning Red States Green, Big-Box Retailers Going Solar, and the Few, the Proud, the Brave, the Green

Individual choice goes a long way toward building the clean-energy marketplace we want. But until utilities, businesses, large nonprofits, and governments at all levels—local, state, and federal—climb aboard, clean energy will continue to trail the field. These supersized entities have too many levers to pull, too much weight to throw around, and too much roof space for solar panels to ignore. In this chapter, we look at what these supersized groups are doing and what you can do to encourage them and make a difference.

One bright day in November 2002, I was walking up the stone steps of City Hall in New Haven, Connecticut, for a meeting with Mayor John DeStefano. The Connecticut Clean Energy Fund had helped support SmartPower's launch earlier that year, and now they wanted us to help Connecticut become a leading clean-energy state.

The Fund, supported by a small surcharge on almost every energy bill in the state, had millions to spend, but at the time there weren't that many initiatives to push. The state had no aggressive solar or energy-efficiency programs—in 2002, hardly the only state in that boat—so we had decided to zero in on Connecticut's new and underutilized green tag program, a scheme that allowed customers to check a box on their energy bill to pay for additional renewable energy to be fed into the power grid. By checking the box, Nutmeggers were helping the whole country's energy supply become a little bit greener.

To get more people excited about green tags, we had created the Connecticut Clean Energy Communities Program and gathered all the relevant stakeholders, a group that included the Clean Water Fund and the Inter-Religious Eco-Justice Network. Together, we instituted a challenge to the 180-plus cities and towns in Connecticut: get at least 100 residential customers in your community to commit to using the clean-energy option on their power bill *and* the town or city government to lock into purchasing 20 percent of its energy from clean sources by 2010, and the program would certify you with bragging rights as a Clean Energy Community.

That was the challenge I was bringing to the New Haven mayor's office that afternoon, but I didn't know what to expect. DeStefano was then forty-seven years old, the son of a New Haven policeman, with a gruff manner. Someone told me he had once been a Green Beret. I was hoping he wouldn't throw me out his office window if he didn't like what I had to say. Thankfully he didn't do that, but he didn't mince words either.

"Absolutely not," he said without even looking up from his desk. "You're crazy. It's going to cost more for us, and no one's asking me to do this. I got to get books in the schools and cops on the streets." When he did look up, his eyebrows arched, he summarized his position: "This is not a priority."

Well, I thought, *at least he didn't jerk us around.*

I tried another angle. "That's fine, Mr. Mayor, but we're trying to build a campaign to give you the support you need to do this. Because what we're hearing you say is that if your citizens want this then you'd be more inclined to do it."

"Have at it," DeStefano said. "If you can get them to tell me they want me to raise their energy rates, then I'll do it." His tone belied his thoughts that it would be easier for a pig to fly first class across the Atlantic.

Not a great beginning, but no time to give up either. The sort of community-based organizing needed to upset the mayor's expectations was where our initiatives were going to live or die, and we already had a good start. Our partners in the Inter-Religious Eco-Justice Network had been appearing at synagogues and churches all around the city. They literally were preaching from the pulpit, asking the members of the groups they visited to commit that church or synagogue to 20 percent clean energy by 2010 and trying to get individuals to sign up for their homes. The Clean Water Fund already had been out canvassing door to door as well, so we asked both groups to encourage people to tell the mayor that they supported clean energy in Connecticut.

After the campaign had been running for about eight months, Mayor DeStefano called.

"Listen," he said. "I'm going to issue an executive order committing the city to 20 percent clean energy by 2010."

Wow, I remember thinking, *He's calling us.*

"And I'll get a hundred customers to sign up," the mayor added.

"We'll make it even easier for you," I said. "If you follow through on this, the Clean Energy Fund will give you a 1-kilowatt solar array for any city building that you want."

"Great, I'll take it," said the suddenly green mayor.

Soon after, we held a big press conference at which Mayor DeStefano signed an executive order doing just what he said—pledging New Haven to 20 percent clean energy by 2010. It was a huge coup. More important, it sent critical signals to the marketplace. Multiple companies—hydro, wind, and solar—sell the clean energy purchased by the green tag program. When a big purchaser like New Haven comes on board and offers long-term contracts, these providers go back to their investors and crow about the deal. That, in turn, encourages more investment in the field, small steps toward the tipping point where selling renewables becomes sustainable and profitable without the help of special programs.

The press conference, though, didn't mark the end of the line, just the beginning. We continued going around town, encouraging people to sign up for the green tag program, but now we had help. Sometimes the mayor himself hit the street with us, urging his constituents to sign up for Connecticut's Clean Energy Option and ensuring that New Haven won its 1-kilowatt solar panel. The city reached one hundred names pretty quickly and became a Clean Energy Community, at which point we upped the ante.

"Mr. Mayor," I offered, "for every 100 additional people that you sign up, the Connecticut Clean Energy Fund will give New Haven another 1-kilowatt solar installation. No limit."

While DeStefano kept pushing the project along, the green tag program was snowballing across the state. New Haven's neighbor, North Haven, wanted in on the program for free solar, too. Soon, cities and towns across the state were pumping the program to their citizens. As the initiative gained momentum, I had another talk with Mayor DeStefano.

"I see that Slifka, that mayor up in West Hartford, is doing this, too," DeStefano said.

"Yeah, we're really excited about it," I said, and we were. With only half the population of New Haven, West Hartford was leading the state in customers signed up for the green tags. New Haven stood second.

"Well, you know what?" DeStefano laughed. "We're going to kick their butts!"

"Can I quote you on that?" I asked, thinking, *This a dream, a clean energy war between two cities!*

"Yeah," said DeStefano, "You tell Slifka that."

"Do you know Mayor Slifka?" I asked.

"Never met him."

"OK," I suggested, "let's sweeten the pot a little more. You challenge West Hartford, and whoever gets the most signups by Earth Day next year wins. The losing mayor has to wear a T-shirt that day saying 'I Wish I Was Mayor of the winning city.'"

Scott Slifka, a twenty-eight-year-old first-term mayor, appeared to have about as much in common with Mayor DeStefano as the wealthy, mostly white suburb of West Hartford did with gritty, diverse New Haven. But soon the two locked themselves into a highly publicized challenge. On August 25, 2005, the two mayors met outside New Haven City Hall to formally announce the competition. "It's not a contact sport, so no one will get hurt," DeStefano said. "But it's like anything in life, it's more interesting when there's a competition."

Then, as a side wager, DeStefano told Slifka that he was so confident that his city would win that he'd treat all the city workers in West Hartford to free slices of New Haven's famous pizza if they lost. West Hartford doesn't really have a local cuisine, so Slifka countered with free soda from a local bottler for New Haven.

For the next seven months, local media in both cities followed the competition up close and personal. Thermometers tracking the cities' progress went up in both city halls. Other state politicians also joined the fray. Though he was closely following the intrastate feud, Middletown's Councilman Ronald Klattenberg wasn't ceding the title either to West Hartford or New Haven. "My position," he told the *New York Times*, "has always been that we'll beat both of them."

He didn't, but everybody won. On Earth Day, April 20, 2006, New Haven and John DeStefano officially claimed victory in the Clean Energy War, and everyone celebrated with local pizza and the soda contributed by West Hartford.

But the story doesn't stop there either. As so often is the case with renewables, once you get people focused, the push takes on a life of its own.

When we kicked off our Connecticut campaign in 2002, none of the state's 180-plus municipalities qualified as Clean Energy Communities—not a single one. Today, about 110 of these local governments are buying over 20 percent clean energy and have a minimum of 100 customers signed up through the Connecticut Clean Energy Options programs. In just eight years, nearly 70 percent of the state's communities bought into renewables. In 2009, Connecticut came from nowhere to pop up at number eight on the National Renewable Energy Labs ratings of the top ten clean energy states, the first time Connecticut had ever even been ranked.

Those two mayors, DeStefano and Slifka, became bosom buddies of clean energy, ultimately running on the same ticket for governor and lieutenant governor of Connecticut. They didn't win, but clean energy didn't lose. The incumbent governor upstaged her vanquished opponents by committing the entire state to 20 percent clean energy by 2010. That's what I call a successful campaign.

But the point is—and it's critical—that this doesn't have to be SmartPower's campaign, or a face-off between two Connecticut mayors. Renewables, going green, doing well by doing good for the planet are contagious ideas. Give them the right push, and they'll go viral on their own, even in places where conventional wisdom might tell you that the odds are stacked in the other direction. Arizona, for instance.

Turning Red States Green

In October 2009, I drove the 200 or so miles from Phoenix to Flagstaff with a magnum of Bailey's Irish Cream in the back seat. I was on my way to meet Don Brandt, CEO of Arizona Public Service, at his vacation home. About an hour into the trip, the terrain surrounding I-17 grew more jagged, the air colder, and any reminders of the balmy desert city where I'd landed had disappeared completely. Suddenly, I was climbing up snow-covered mountains.

When I finally reached Brandt's house, tucked into the hills outside Flagstaff, it offered a welcome relief, warm and inviting, not ostentatious in the least. There was also that huge bottle to soothe my nerves. I had heard that Brandt liked Bailey's, but I had hoped to find a slightly more manageable size.

As we sat on the sofa chatting and having a few drinks, it occurred to me that Brandt is exactly what Arizona and renewable energy needs: someone with smarts and long-range vision but, just as important, a normal guy. Most utility CEOs fit into a narrow type: a businessman who is all business. In a suit, wearing almost invisible wire-frame glasses, and with a sheen of silver in his hair, Brandt can look the part, but his easy laugh, the twang in his speech, and his laid-back personality make him much more approachable. He's in the business of energy, but he's got a larger vision for his state, and with good cause.

Ever since it was a scrappy federal territory, Arizona's fortunes have risen and fallen with its amazing *im*balance of natural resources. The state produces two-thirds of the nation's copper but is also one of America's most water-hungry spots, largely dependent on snowmelt from mountain states to the north. At the beginning of the twentieth century, with settlers in the new town of Phoenix frequently facing drought followed by flooding, the territory gladly accepted the federal government's offer to build a dam across the Salt River. The result—Roosevelt Dam, the world's largest masonry dam—created a reliable water supply near Phoenix. Twenty years later, the structure was converted to produce hydroelectric power.

Since then, Arizona has been one of the fastest-growing states in the country and one with enormous energy needs. Much of the state's population lives in desert. As a result, just about everyone uses air conditioning and swimming pools—often the second biggest power suck in an Arizona house—year-round. To meet those growing needs, Brandt's company, Arizona Public Service, brought the nation's largest nuclear generator online in the 1980s. APS also runs coal-fired plants. But Brandt wants to make renewables an increasingly larger part of the state's energy portfolio, and I'd come, bearing Bailey's, to help.

On the one hand, solar power in Arizona is an absolute no-brainer. It's a desert! They have sunlight like Houston has barbecue. Where else are you going to sell solar? But this *real* sunshine state, with Tucson, Phoenix, Yuma, and Flagstaff all among the ten sunniest cities in America, is also the land of Barry Goldwater, dominated by the kind of conservative politics that puts a burr under the saddle of renewable energy initiatives. Arizona reliably sends two Republicans to the Senate as well as a house delegation dominated by Republicans. Only one Democratic presidential candidate—Bill Clinton in 1996—has carried the state since Harry Truman. Forget sunny weather. This is a political environment designed to kill renewable-energy initiatives dead. Indeed, some politicians in Arizona point-blank don't like solar power and consistently try to do just that. Just the idea upsets them. So Arizona ironically combines the best and the worst of renewable energy's possibilities.

The state's current governor, Jan Brewer, boasts serious conservative credentials of her own. She's pro-life, signed a bill allowing guns in bars, and is best known for supporting the toughest anti-immigrant law in the country. She's no crunchy liberal environmentalist. But under her staunchly right-wing watch, which began ten months before I visited Don Brandt, the state's utilities (under the direction of the state utility commission) offered huge incentives to spur residential solar installation. In a time of extreme austerity, when cutting taxes has become a mantra, a surtax on power bills has actually expanded renewable energy there. Far from being the Deadwood Gulch of renewables, Arizona has seen a tremendous growth in solar over the past few years. In a state with a huge budget deficit, Governor Brewer is both more than happy and completely able to tout the growth of solar jobs that has made her state number three in the nation.

Politics aside, as they should be, Arizona's emergence as a solar state comes from the same common-sense approach toward electricity that Don Brandt has. When you need as much energy as Arizona, you can't really care where it comes from, and you can't assume there will always be enough. The state nearly had to institute extreme water rationing a few years ago, so, with as many resources as the state consumes, it will happily take out an insurance policy.

This agnostic approach to electricity generation isn't unique to Arizona, either. You can find it all around the country, even in traditional energy states like Pennsylvania and Texas, the home of Big Oil. In fact, it might be precisely *because* these are historic energy states that they understand renewables so well. When you make your living from energy, you have to keep looking ahead to where it's going to be ten and twenty years down the road. But Arizona is the story most worth telling, and Don Brandt's visionary yet pragmatic approach to power generation has a lot to do with it.

Since he took over as CEO at Arizona Public Service in 2009, Brandt has developed the plan for Solana (the Spanish word for solarium), one of the most ambitious solar-energy-generation projects in the world, set for completion in 2013. Located just outside Gila Bend, a desert town of a few thousand once known as the Crossroads of the Southwest, Solana will soon produce enough electricity to power 70,000 homes in Arizona at full capacity, an amazing and entirely replicable demonstration that solar is a viable, scalable energy source.

One of the largest solar power plants in the world, Solana will provide APS with more solar electricity per customer than any utility in the country.

Aside from its sheer enormity, Solana uses a unique technology called concentrated solar power. Instead of photovoltaic cells or solar water systems, concentrated solar power uses a huge field of parabolic mirrors to focus the sun's rays on pipes carrying a heat transfer fluid that can reach a temperature of 735°F. When returned to generators, the fluid transfers its heat to water, producing steam, moving a turbine, producing DC current, and releasing zero greenhouse gasses. This heat can also be stored in a molten-salt mixture to generate electricity from solar even after the sun sets. Imagine flipping on your lights at 11 p.m. and realizing that they're running on the sun that set hours earlier.

Solana is just one of the game-changing strides that Arizona Public Service is taking to shape America's clean-energy future. They're also creating the Community Power Project in Flagstaff, which offers residents a flat utility rate for twenty years in return for letting APS put solar panels on their roofs. (The electricity generated feeds back into the grid.) The 300 homes that are a part of the program will, according to APS.com, form "in essence, an interconnected renewable power plant." The Flagstaff network also will feed into the smart-grid system being developed by APS, which will help revolutionize the tracking of energy use and efficiency.

The most encouraging part of Brandt's solar strategy is not that it's good for the environment, which it is, or that it's the right thing to do, which it also is. He's using hard-line business practices and strategies to do what's best for his company. Solar makes business sense for APS, just as it makes business sense for Arizona *and* for all those Arizonans who are getting solar panels on their roof and locking in energy rates two decades out.

Impressive clean-energy initiatives abound all across Arizona, spurred by a state mandate that utilities produce 15 percent of their energy via renewable resources by 2025. Peer out your airplane window on the final approach to Tucson International, for example, and you'll see an expanse of reddish earth dotted with low-lying scrubby growth, snaking subdivisions, and, just south of the airport's runways, a huge solar array sucking up the sun's plentiful energy. That massive photovoltaic installation can produce up to 5 megawatts of power, an amount estimated to meet the needs of about 1,000 homes. Pilots initially had expressed concern that light bouncing off the panels would blind them, but experience has long since laid those concerns to rest.

To the west of Tucson, Yuma International Airport has introduced its own initiative in alliance with APS: a $4 million solar shade over the parking lot that keeps everyone underneath cooler while sending electricity back to the passenger terminal. (The University of Illinois's entry in the 2011 Solar Decathlon used the same technology.) That shade should nearly flatten the airport's electric bill increases to less than one half of 1 percent a year for the next 20 years.

Just as Arizona overcame potential political storms to become a leader in solar technology and job creation, so the state has become an unlikely prime test market for electric cars. Sure, environmentalists are excited about the machines' lower CO_2 emissions, but utilities like APS also see awesome possibilities. To power companies autos are just another appliance, a really big, really hungry appliance that frequently recharges during nonpeak hours. No wonder they're hustling to build recharging stations all around Arizona. If you create the infrastructure needed to make plug-in cars viable, this fledgling market sector has a lot better chance of prospering, and when that happens, we all come out better.

Bottom line: Power providers don't have to be the enemy. There's plenty of common cause to be explored, and if America gets greener, who cares why?

The City of Roses

About 1,300 miles northwest of Phoenix lies Portland, Oregon. Not only are the two cities far apart physically—Phoenix in a desert about three hours from Mexico, while Portland sits in a river valley one state away from British Columbia—they seem to exist in two different climate universes. Phoenix boasts 300 sunny days a year, while Portland is the fifth rainiest city in the country, with 155 days of rain or snow each year. Because Portland sits on the 45th parallel, the same as Ottawa and Montreal, its residents trade long summer daylight hours for very brief ones in winter, with just ten hours between sunup and sundown. So, just as solar in Arizona seems obvious, covering the moss-clad roofs of western Oregon with photovoltaic panels is a waste of time. Right?

Not so fast.

Though Portland's weather and latitude mean the city gets only about 60 percent of the amount of useable sunlight as Phoenix, solar works beyond the Sunbelt. Oregon in general and Portland in particular prove it. They're not letting an excess of rain clouds get in the way.

Over the past few years, Oregon has boosted its solar stature with incentives leading to a connected capacity—the amount of solar electricity the state pumps into the power grid each year—that has nearly tripled from 7.7 megawatts in 2008 to 23.9 in 2010. While the state still distantly trails leaders like California and Arizona, its capacity is about 70 percent of that of Texas, a sun-rich state more than twice Oregon's size. Oregon has also made solar an important part of the state's infrastructure. In August 2011, the Oregon Department of Transportation broke ground on the world's most powerful solar highway, a photovoltaic array placed next to an existing freeway that will produce about 2 million kilowatt-hours a year, power used partially to offset the cost of highway lights. The state also launched tax and other incentives to lure solar companies here, efforts that have led Solar World, a German company, to open the largest solar cell manufacturing facility in North America just outside of Portland.

Just as the Clean Energy War between New Haven and West Hartford spurred Connecticut to become a clean energy leader, so an intense national competition to host solar companies, with the hope of becoming the "Solar Silicon Valley," has become especially heated among Oregon, California, Arizona, and Texas, but also East Coast states including New Jersey.

Yes, New Jersey.

A new Solar Energy Industries Association report confirms that New Jersey has surpassed California as the state with the largest commercial solar market. Photovoltaic installations in the Garden State now account for 24 percent of all US installations, showing an unparalleled growth rate of 170 percent between the first and second quarters of 2011. New Jersey's new standing also includes massive solar projects such as the Gloucester City Marine Terminal, which will produce enough energy to power 1,500 homes. The number of jobs created from these projects is also no small sum. New Jersey's initiatives have created 5,000 solar jobs, not including indirect solar-related work for bankers, architects, engineers, roofers, landscapers, and more.

Nor do those job figures take into account small businesses across the state that have piggybacked on solar to rationalize their energy costs. Sharrott Winery in Winslow, at the southern end of the state, now makes and chills about 3,500 cases annually of locally produced wine, using a double row of free-standing solar panels at the rear of its fifty-eight-acre tract. Pour a glass of that, and here's to proof that every state can be a clean-energy state.

Let the Winds Blow

A few years back, the *New York Times* ran a feature about the small town of Ainsworth, Nebraska, population 1,728, and its unique annual event, the Middle of Nowhere Festival. Ainsworth lies in the north-central part of the state, near a geological phenomenon known as the Sand Hills. That's great if you're a birder. Sand Hill cranes are beauties to behold, but it doesn't do much for the soil when it comes to growing crops.

Ainsworth does, however, have one notable natural resource going for it: wind. Average speeds in the area run to almost 20 mph. During the Dust Bowl days of the 1930s, this was the wind that stripped fields of their best soil and drove farmers off the land. Today, it's producing the finest crop Ainsworth has ever grown: wind energy.

Built by the Nebraska Public Power District and opened in 2006, the Ainsworth wind farm sent dozens of windmills more than 200 feet up into the big Nebraska sky. They're so enormous that they became an instant tourist attraction, even as they provided power for about 19,000 homes. That's more homes, by a factor of ten, than exist in Ainsworth, or even Brown County, of which Ainsworth is the seat. And none of that power goes directly to local residents in any event. With an operation this big, the national grid gets the juice. Ainsworth's production goes to local utilities and distant ones as far away as Jacksonville, Florida.

Clean, inexhaustible energy for nearly 20,000 homes is tremendous progress. But the larger story is the attitude within the community toward the giant windmills. A *New York Times* article on the wind farm noted that, while at first they were seen as exciting and different, now they're simply no big deal. That I find not just interesting, but amazing.

To the people around Ainsworth, Nebraska, clean energy long ago ceased being an argument between greenhouse-gas decriers and greenhouse-gas deniers. It's not futuristic or political. The clean energy in question is in their back yards.

It's dependable and available. Accessing it doesn't require blowing up mountains or laying pipelines through fragile terrain or injecting high-pressure solvents deep underground. Wind is already there, safe for wildlife and humans.

The Ainsworth turbines underwent an acid test even before the wind farm opened. A November 2005 blizzard that swept across the Great Plains produced wind gusts up to 114 miles per hour, with average speeds at the wind farm measuring 83 mph. Those are hurricane conditions, but nothing gave. Everything held. Will that always be the case with wind turbines? Obviously not. Machines have flaws, and weather can never be 100 percent predicted. But wouldn't you rather have a wind turbine disintegrate than a nuclear-power plant? That's not even a close call.

"How Can I Help You?"

If you're one of the 84 percent of Americans who visit a Wal-Mart at least once a year, you've probably noticed the stores' endless aisles and "How Can I Help You?" emblazoned on the back of the blue vests worn by the ubiquitous staff. But the world's largest retailer has been making some changes in its stores.

Beginning in 2007, prototype stores began popping up around the country, using natural sunlight to brighten the stores and LED lights in refrigerator displays that automatically dim until customers approach. Even harder, in fact, almost impossible to notice have been the solar arrays raised on scores of the giant retailer's enormous roofs over the past several years.

The roots of this retail empire's foray into green power date back to 2005 when the company announced the incredibly ambitious and completely vague goal of being 100-percent reliant on renewable energy sources. As the country's top private consumer of energy, such a switch would have enormous impact, but the lack of any accompanying timeframe didn't leave many people holding their breath for immediate results. Over the past several years, though, Wal-Mart has made significant steps in that direction, continuing to install solar panels and some wind turbines on dozens of their stores' roofs, primarily in California and Arizona. Lefties might have a hard time swallowing it—like Richard Nixon's 1972 trip to China—but the evidence is pretty irrefutable. Wal-Mart is going renewable. *The* big-box retailer is painting itself green.

As remarkable as that seems, Wal-Mart's experiments with renewables actually pale in comparison with a few other retailers such as Kohl's, which recently reported that just under half of the company's roughly 1,300 stores had an Energy Star designation from the EPA. That designation doesn't come lightly. To qualify as an EPA-certified Energy Star site, a business has to jump through a number of hoops. But with more than a hundred locations from California to Maryland featuring solar, Kohl's also claims to be the nation's largest retail host of solar power. Even more surprising, the department store buys enough renewable energy certificates—like those green tags in Connecticut—to offset 100 percent of its energy use. That means that while Kohl's has no way of knowing the source of the energy it draws off the grid, it's offsetting all of it with clean, renewable energy certificates.

What happened? Has corporate America suddenly discovered its conscience? That might be part of it, and maybe it's more good marketing. No company wants its brands tied to despoiling the environment. But that's the small story behind the bigger story, and Wal-Mart shows why. The company's move into solar falls in line with its famous business model of squeezing every last cent out of costs, from suppliers onward to the moment a purchase is made and a product leaves the store. In other words, laudable as these green efforts are, they're not a PR move so much as an extension of Sam Walton's legendary penny-pinching.

And why not? Well before the company began installing solar on roofs, Wal-Mart was carefully monitoring total energy use back at corporate HQ in Bentonville, Arkansas. Wal-Mart rearranged its convoys to save fuel costs. It saved energy by harnessing water to heat and cool its buildings as well as improving its refrigeration system. Maybe instead of asking "Is Wal-Mart sincere about solar?," the question should be "Why wouldn't a company already obsessed with energy expenses embrace renewable energy?" The company's embrace of solar and wind has more to do with the John D. Rockefeller School of Efficiency than with trying to save the planet.

Yes, before Sam Walton, another plutocrat was renowned for his frugal dress and habits, a cheapskate who sometimes fixed the pipes and joints himself at his fantastically lucrative Standard Oil refineries, and further saved bucks by helping his employees make their own barrels. Fossil fuels made John D. Rockefeller the Bill Gates of his age. If he were alive today, he'd take a hard look at solar, too, just like Sam Walton did and for the same reasons. Wal-Mart's move into renewables isn't a loss leader, either. All of the projects must provide energy more cheaply than traditional sources like coal and oil. So far, Wal-Mart's initiatives have saved more than $1 million, a drop in the bucket for a company with annual revenues of around $400 billion, but a measurable savings nonetheless. Those savings will increase dramatically once the company outfits all its stores, not just the prototypes, with solar panels and wind turbines, and Wal-Mart knows it. What better endorsement exists for renewable energy than the embrace of a company infamous for being ferociously bottom-line driven?

Maybe that's why companies that go green so often outperform their counterparts. They understand better than most what really drives the bottom line. In 2011, *Newsweek* released numbers showing that, over the previous two years, companies ranked in the magazine's top 100 green companies for 2009 had outperformed the S&P 500 by 4.8 percent. Together, these corporate environmental leaders were up 15.2 percent, compared to 10.4 percent for the S&P 500. The stock of the magazine's top-ranked green company, IBM, rose more than 100 percent over the past five years, a period during which the S&P 500 overall was flat.

While the top twenty companies in the *Newsweek* green rankings come largely from the technology and health-care sectors, the greenest corporate leaders from other sectors have not only survived the financial woes better than their direct competitors; they've also been emerging from the morass even healthier. Ford, a company on the brink of collapse a few years ago, has released a suite of efficient vehicles much more attractive to consumers facing higher gasoline prices. The automaker's stock price has risen over 50 percent in the last two years. Notice a trend?

Wal-Mart doesn't even appear among these top-ranked companies, but it has achieved some of its savings with techniques that every company with an empty roof could use. Installing solar film technology—sometimes called foil because you can roll it out directly onto an existing roof—is one such technique. The foil is less efficient than traditional solar panels but requires fewer raw materials and can be more suitable for large rooftops. The foil is also cheaper, and for Wal-Mart it's free. But this isn't a sweetheart deal that the world's biggest retailer muscled out of a supplier.

Through what's known as a power purchase agreement (PPA), Wal-Mart allows a third-party company to install solar—foil or more traditional panels—on its roofs in exchange for locking in power rates for a fixed amount of time. The installer owns the energy created, but Wal-Mart incurs no installation or maintenance costs and can accurately predict electricity costs for as long as the arrays are on its roof. This clever hedge against rising power expenses helps the retail giant control a factor over which it has little control otherwise. Such PPA arrangements are a great deal for everyone with a roof—it's just sitting there not doing anything else, after all—and more and more businesses are tapping into the unrealized resource right over their heads. In fact, the biggest problem PPA outfits face today is that there's too much demand.

Want to see a rooftop installation at work? Go to www.american.edu/finance/sustainability/real-time-solar.cfm to track, in real time, the behavior of the first solar installation at American University, in Washington, DC. The rooftop array went live in July 2010 on AU's new School of International Service building. You can also see the array's carbon offset. In the first sixteen months of operation, it generated enough electricity to meet the daily needs of almost 1,600 houses, offsetting over fifty-five tons of carbon. The School of International Service is just one flat-top building in a city with thousands. Every one of them—hotels, office and apartment buildings—has the same potential.

Why American University? Well, its smart administrators understand the bottom line as well as Sam Walton. But more important, AU knows that clean energy is a powerful marketing tool. Indeed, as one top administrator told me, the three questions that college recruiters most commonly hear are: (1) What are the dorms like? (2) How's the food? (3) What's your sustainability policy?

Getting the best and brightest to attend your institution of higher learning is tough competition. AU has learned that nothing says sustainability better than putting renewable energy and energy efficiency into action. Maybe you would send that message to your alma mater. Or if your workplace has available flat-roof space, you could spread the word about PPAs. As we've seen, businesses are happy to go green once they see the bottom-line benefits.

"Ten-Hut!"

Wal-Mart may be the country's biggest retail consumer of energy, but the Department of Defense leaps ahead in total consumption. Our military's energy bill is what you might expect of an organization that provides a place to work, and often live, for two million people across a worldwide network of bases covering territory roughly equal to the state of Pennsylvania. The nature of the work is where the real costs come in. Jet fuel, for example, is a particularly large part of total consumption.

But sometimes, the bigger they come, the harder they try. The Department of Defense set a goal of powering the military by 25 percent renewables in 2025, and its sheer size has made it a leader in reducing reliance on fossil and nuclear power. With so much property under its control, the US Navy has suddenly become the country's largest installer of solar.

In the last chapter, you saw ways of reducing the enormous vampire load in military housing through changing personal behavior. We also learned that the Pentagon was using the old carrot-and-stick technique to push the process along by putting military families on notice that Uncle Sam wouldn't pick up the utility tab forever, while simultaneously upping rooftop-solar installations on private military residences. But because of all the different environments in which the military operates, it has also had to adapt and create its own specialized systems.

In August 2006, Major General Richard Zilmer of the US Marines was angry. The ranking commander in western Iraq at the time, including the deadly, Sunni-held Anbar

Province, Zilmer was sick of seeing his supply convoys attacked as they drove across baking, wide-open desert highways. The attacks were destroying supplies, increasing operational costs, and wounding and killing the Marines deployed to drive or defend these missions.

Zilmer already had armed protection riding along with the convoys. He tried running them at night. But there were too many convoys to slow down the attacks. He could reduce the number of convoys, but everything onboard had a vital importance: subsistence supplies, ammunition, and what Zilmer described as a "preponderance" of fuel. Iraq had basically no reliable energy infrastructure. The diesel-powered electric generators that ran everything, including Marine bases, littered the country.

Unable to do without food or bullets, Zilmer requested a new third leg of the triangle: renewable-energy generators. We're out here baking in a desert, he figured. Why don't we have solar panels? There's wind here, too. What about some turbines? Any or all of them would help us keep down fuel expenses and reduce the number of convoys. Remember, Zilmer was making this request from inside the country with the fifth-largest oil reserves in the world. But reducing his troops' dependency on fossil fuels meant saving lives. Zilmer submitted his request as an urgent priority-one need and suggested a currently existing, American-made technology. But the wheels of military bureaucracy churned and churned, and nothing happened.

Jump ahead two years. Zilmer found himself now in Afghanistan—different country, same problem. US military bases were operating in hostile territory accessible only by winding through small roads in remote areas. The soldiers were still prime targets for roadside bombs, and the convoys supplying them with fuel were routinely attacked. This time, though, Zilmer got his way. His request for renewable energy generators was approved in 2008. Nine months later, the new technology began to arrive, including a solar-and-battery combination transportable in Humvees that could still power a combat outpost's gear while withstanding tough field conditions.

That was just the beginning of bringing renewables to the battlefield. In the spring of 2011, US Marines began conducting three-week patrols through one of the deadliest zones in southern Afghanistan, carrying rollable, mini-solar panels to power their battle gear. These solar mats, about the size of a hand towel, weren't being forced on the Marines by scientists back in American laboratories. They were specifically constructed to meet a field demand.

It turns out that the modern American soldier has some of the same problems as our modern houses. Our houses are infinitely better sealed and smarter than the drafty places where we grew up. But all the new gizmos—flat-screens, laptops, smartphone chargers—totally overwhelm our upgrades in efficiency. Similarly, modern soldiers are more powerful, lethal, and efficient than any previous force, but the same technology that makes them so deadly also makes them energy sucks. In our homes, it's a flat-screen. In Afghanistan, it might be a battle-tough navigational computer or night-vision goggles. Either way, today's soldier uses about four times as much power as his counterpart used just fifteen years ago.

Most soldiers walk around with hundred-pound packs, but the heaviest items inside are electrical gear and batteries. In both cases, solar provides an important part of the solution. The green technology carried by many battlefield Marines today makes their packs about twenty pounds lighter and allows them to recharge their laptops, radios, and GPS devices without lugging heavy batteries on patrol. Fuel use also has dropped at bases where solar has been installed. In fact, some bases rely completely on solar.

But Marines aren't using solar just to increase their military advantage. They're also looking at renewable technology as a way to increase stability in a country with a devastated, almost nonexistent infrastructure. Yes, there are unusual barriers to installing such devices in an almost-lawless country. Photovoltaics on the ground had to be moved to roofs after being stolen and vandalized, while tall wind turbines made inviting targets for terrorists whose main goal is to destabilize the country any way possible. But the successes offer incredible hope.

The Afghanistan National Security University, a campus expected to number around one hundred buildings and considered critical to insuring the country's stability once American forces leave, has been selected as a test case. Not connected to any electrical transmission lines, the university is powered primarily by diesel generators. Because its high walls make the site relatively secure, US planners are pushing enthusiastically for wind and solar installations to make the training center as self-reliant as possible, reducing the same need for vulnerable supply lines that American military planners have tried to circumvent at their forward bases.

Solar also has proven valuable in the chaotic world outside the gates of the university. Kabul, for example, the Afghan capital, has no reliable electricity grid for delivering power. So the US military teamed up with the mayor's office to install about thirty poles, each topped with two lights and used to illuminate shopping districts. The self-contained street lamps collect the sun's rays during the day to power low-wattage LEDs at night. Technologies like these are more than convenient and more than clean-energy success stories. They can help hasten a successful American exit from a more stable, more prosperous Afghanistan. As Army Corps of Engineers Colonel Thomas Magness told *Climate Wire*, "It's not about green technology as much as it is about sustainable solutions for electricity and power here in Afghanistan."

Just as Wal-Mart's clean-energy push has less to do with environmentalism than with plain-old-business sense, the DOD's embrace of renewable energy is happening within the context of a completely non-environmentalist core mission, the protection of American personnel. As Secretary of the Navy Ray Mabus put it, lowering carbon emissions "is a good by-product, but it's a by-product."

WHAT YOU CAN DO

BASIC:

- If you shop at Wal-Mart, find out if your location is a prototype store that has a solar or wind installation. If it does, find the manager and say thanks. If it doesn't, find the manager and ask why not.
- How about your college, or your children's school, either present or prospective? What's its sustainability policy, and what role does clean energy play in it?

INTERMEDIATE:

- Contact your mayor/city manager/county commissioner and urge him or her to make your community a clean energy leader.
- Find out if your church, synagogue, mosque, or other place of worship is using renewable energy. If not, help lead the charge.

ADVANCED:

- If your business owns a flat-top building in which it operates, spread the word about solar PPAs up the chain of command. Leading by example makes good business sense.

5 OVER THERE

Wide World of Renewables, Energy Profiles and Energy Policy, American Exceptionalism

In the arid southwest corner of Spain, Phoenicians, Romans, Arabs, and Vandals in turn came and went over the past several millenniums, each culture leaving behind its distinctive architecture, language, customs, and tradition. Legend has it that Hercules founded Seville, but in 2007 the city became the site of a new project—a completely modern labor.

Just fifteen miles outside the city's historic center, a field of 624 large, moveable mirrors, known as heliostats, fix their gaze on a 380-foot tower where a solar receiver sits. Focusing the southern Iberian sun's rays on a huge container of water boils off steam that drives a turbine. The turbine's power transfers to a generator, creating electricity converted into AC current and pumped into Seville's electrical grid, powering everything from air conditioners to espresso machines. The 11-megawatt plant was Europe's first commercial concentrating-solar-power plant—and a prototype in the development of the solar thermal technology in Arizona that heats a molten saltwater mixture to generate power even after the sun sets.

The Seville project is just one of dozens embraced by Spain in recent years, making the country one of the world's pace-setting solar venues. By 2010, Spain was producing the third-highest rate of solar per capita, and its solar companies were winning contracts globally. But the land of Don Quixote hasn't given up on windmills. Spain has the fourth-highest installed-wind capacity in the world and the highest per capita.

Next door, Portugal also has made huge strides on a variety of energy initiatives over the past several years. Without Spain's size or the resources of wealthier nations to the north, Portugal has cobbled together a frugal combination of wind farms, hydroelectric, and solar, and experiments with wave power to make its goal of producing 60 percent of its electricity from renewables by 2020 a realistic one.

For this relatively poor member of the European Union, generating this much electricity from renewables once seemed highly unlikely, but that's really in many ways the point. Arizona, which might lead the US in imaginative renewable projects, is far from America's wealthiest state. California, another renewables leader, teeters on the verge of bankruptcy, much as Portugal itself does. Yet as Portuguese Prime Minister José Sócrates says, "The experience of Portugal shows that it is possible to make these changes in a very short time."

Fire, Ice, and a Big Red Snake

In the high latitudes of the Atlantic sits an island born of volcanic eruptions at the meeting place of the Eurasian and North American plates. Essentially a cooled plane of lava plopped in the middle of the ocean, Iceland isn't naturally a good land for agriculture. It has no oil or coal reserves and very few natural resources of any kind, except the fish offshore and a national genome map second to none. But this tiny island nation has tapped into the massive reservoirs of hydrothermal energy roiling up from deep within the earth to produce all of its electricity domestically. That's right. *All.*

Though Iceland still imports everything from Coca-Cola to cars, effective use of its energy resources means that Iceland is impressively independent in electricity generation. It imports petroleum for cars, but it has recently become a testing zone for electric-vehicle recharging stations. Once an effective infrastructure exists, Iceland will still have to import the vehicles, but domestic steam will power them. Is the Iceland example on the mind of energy start-up AltaRock and its big-money backers at Google as they work to exploit the rich geothermal potential of the Pacific Northwest? You bet. That's why AltaRock's Jim Turner calls geothermal a "pot of gold."

East of Iceland, a 600-foot-long red snake writhes in the water off the coast of Scotland's Orkney Islands. Like Iceland, Scotland's latitude makes solar less effective, so it has become a world leader in experimenting with machines like this red snake that converts the energy of wave motion into electricity via hydraulic cylinders. Generated power feeds back to the Scottish mainland via an undersea cable. The machine, called a P2 device, can generate 750 kilowatts of electricity, enough power to supply approximately 500 homes throughout the year. That's a tiny fraction of Scotland's houses and needs, but it's surrounded on three sides by water and waves. Maybe Nantucket, which is surrounded on four sides by water and waves, should be filling the ocean with snakes rather than filling the distant horizon with wind turbines.

Heating Up

Further east at just about the same latitude as the Orkney Islands sits Stockholm's central train station, a building that now taps into a renewable energy source as novel as it is omnipresent. As you exit the train at Stockholm central, your body itself becomes a renewable energy source.

Built in the 1870s, Stockholm's main train station—like many major stations built in the West at the time—has high, vaulted ceilings. In the middle of cold, dark winters, this vast space generates enormous heating bills. So several Swedish engineers decided to harvest the heat from the roughly 250,000 people that pass through the station daily. At busy times of the day, heat exchangers in the ventilation system capture the heat rising off commuters' bodies and funnel it into large underground tanks of water. Once this water warms up, you've got a surprisingly potent heating source. This basic idea is already in use in a few other facilities, but the engineers in Stockholm took the technique one step further, pumping the water through a nearby office block, reducing that building's heating costs by 20 percent—all with free heat rising off passengers waiting for trains next door.

What's to keep something like that from happening at New York's Grand Central Station? An estimated 750,000 people pass through Grand Central daily—triple the Stockholm number—and over a million on holidays. Imagine Grand Central heating the Met Life building next door. In fact, only a lack of imagination stands in the way of something like that actually happening.

In North Africa, renewables are partying just as hard in some places. In the Moroccan desert, south of the Atlas Mountains, lies Ouarzazate, the nearest town to the future site of a 500-megawatt solar thermal site expected to begin construction in 2012. The project will take up over four square miles of the sparsely inhabited, dusty red land as the first African effort of Desertec, an enormously ambitious network of varied renewable-energy sources linking Europe, North Africa, and the Middle East.

Desertec plans to capture renewable energy from wind farms on the west coast of Africa, solar installations throughout the Sahara and Saudi Arabia, and hydropower from the Nile and route it back to Europe through a grid installed under the Mediterranean. Desertec's European backers hope that as more such projects develop it can provide 15 percent of Europe's entire energy needs by 2050. This isn't just a one-way stream, though. Although a variety of European utilities, banks, and companies with expertise in renewables—Deutsch Bank and Siemens among them—are providing financing, construction, and technical assistance, the host countries have first dibs on Desertec's power.

Overall vision exceeds the political and practical realities of the moment, but the opportunity is alluring enough that several other North African countries are negotiating to join the network. Tunisia is discussing building a solar farm, and Algeria is a favored entrant because of its proximity to Europe. Egypt, meanwhile, is planning a 1,000-megawatt wind farm in the Gulf of Suez, to be operational by 2016. Countries further east, including Turkey, Syria, and Saudi Arabia, will have opportunities to join as high-voltage, direct-current cables continue to extend across the region by 2020.

Let's remember what region we're talking about here. This is the heart of OPEC, a group whose fortunes have come, for better or worse, from the discovery of vast reservoirs of fossil fuel in the past hundred years. Now, through Desertec, solar-concentrating plants on the Arabian Peninsula may provide Europe with energy independence from imported petroleum products. That this is even being discussed is revolutionary. Shouldn't someone be asking if North America's four largest deserts—the Chihuahuan, Sonoran, Great Basin, and Mojave—want to join the party or start their own?

Then there are the really wild and crazy solar moments around the world—Vauban, for example. A former World War II military barracks just outside Freiburg in southwest Germany, Vauban was occupied by the French until the end of the Cold War. When they left in 1992, the base fell into disuse except for some hippie and anarchist squatters. In the mid-1990s, the Freiburg City Council accepted the request of a group called Forum Vauban that petitioned to develop the site in an eco-friendly way. The result is a suburb in which seven in ten residents don't own cars—even though the majority commute to work—plus the Sun Ship, the world's first energy-positive commercial building, and the Solar Settlement, a fifty-nine-home community in which all the houses produce a positive-energy balance. In fact, *every* home in the whole development produces an energy surplus that Forum Vauban sells back to the city's grid for per-home profits averaging around €4,000 (roughly $5,300) per year.

In fact, Vauban sounds a good deal like Portland, Oregon, on steroids, but it could also be an inspiration to a place like Detroit trying to remake itself after decades of decline. Green is good for urban redevelopment, too.

Japan's Kasai Energy Park, Sanyo's experimental office "farm," lies just outside a small town mostly known for its five hundred stone statues of the disciples of Buddha and the world's largest globe clock. After passing small farms and rice paddies along the side of the road, park visitors come face to face with a sleek, glass, multistory administration building wrapped in solar panels—and not just on top but around the building's southern façade as well. Inside, a circular glass "solar table" contains a photovoltaic panel that, when placed near a window, wirelessly recharges handheld devices. There are recharging stations for electric bikes and cars. Smart-grid technology uses sensors and visualizations of energy targets to reduce CO_2 emissions at the Kasai park by 2,480 tons every year, equal to the amount absorbed by over 175,000 cedar trees. Imagine an American college campus that could begin to make the same boasts. The best and the brightest applicants, the most tech-savvy and green-conscious, would be beating down its door to get in.

To create an icon for Rio de Janeiro's 2016 Olympics as environmentally engaged as the country itself, Brazil, always on the edge of everything these days, is developing a massive outdoor structure made of two main elements: a ground floor base that juts over the edge of a small island that contains an amphitheater, auditorium, cafeteria, and shops topped by a huge sail-like slab. The building is designed to be powered entirely by solar panels. Excess power during the day will pump seawater high up a tower to power turbines that create energy during the night. On certain occasions—and no doubt the Olympics will qualify—the water will cascade down the building's flat face and into the ocean as an artificial waterfall. According to the architecture firm that imagined the building, it is a striking example of structures that are both beautiful and preparative for the "imminent post-oil future" through use of sustainable energy—a gutsy statement given Brazil's status as a major oil producer and exporter. But in a larger sense, gutsiness lies at the heart of many of the initiatives we've just seen: the courage to imagine the future and do something about it. That's as true for the nations of the world as it is for each and every one of us who live in them—Moroccan, Brazilian, American, or anyone in between or beyond.

In (and out of) Hot Water

One final example from beyond America's shores illustrates both how similar the US is to the rest of the world when it comes to renewables, and how different.

Over the past several years, China has spent tens of billions on renewable initiatives, well ahead of America (number two) and the rest of the world. China has the highest installed-wind capacity in the world and is the largest hydroelectric producer. Solar water heaters are ubiquitous in the country, and China leads the world in production of solar panels—even if it doesn't particularly consume photovoltaics yet.

Those stunning numbers come with caveats, though. China's investment in renewables is impressive, but forcibly evacuating huge numbers of people to create floodplains and ignoring environmental effects when building dams means that the Chinese experience has very little in common with our own. American companies would face enormous fines and harsh public scrutiny if they simply dumped the large amount of dangerous chemicals involved in producing solar panels, as China does. State-driven initiatives simply don't match up well with the dictates of free-market capitalism.

As in China, virtually all residential hot water in Los Angeles was solar-heated once upon a time. But our communities have long been wired for electricity and piped for gas. We can pick water heaters from a wide variety of styles, efficiencies, and sizes, from huge tanks to tiny on-demand systems. If our awareness of the solar-water option continues to grow, an increasing number of us might well choose to install one next time our water heater breaks. But for many Chinese, this is the only option—the hottest water they'll ever get. Of course, the Chinese lack of choice has positive environmental ramifications, but if we're trying to expand the use of renewable energy in a wealthy consumer society, we need to assume that we will have choices, and that renewables have to compete on the basis of the value decisions we all make every time we go to the store or shop online.

America is not China, Portugal, Sweden, Tunisia, or anyplace else. We are a specific spot on the planet with a specific history, energy profile, needs, and ways of getting where we need to be. The federal government isn't suddenly going to mandate that 50 percent of energy come from wind by 2020 and pour massive resources into this effort. We can't even count on a new federal energy bill mandating moderate increases in sustainable energy or even a congressional consensus on global climate change. But as we've seen, the future of solar won't come from Washington loan guarantees or a climate bill. It will come instead from millions of us American consumers educating ourselves on what the problems and possibilities are where clean energy is concerned, and then wanting to buy clean, renewable energy and be energy efficient. That's a good thing.

What we do best in America is identify a market, innovate and create products to satisfy that market, attract entrepreneurs and fresh capital, improve those products, and use those improvements to open new markets. That's how capitalism works, and that's where we go next: how to bring basic marketing principles to what often seems a holy cause.

WHAT YOU CAN DO

BASIC:

- If your head has been in the sand, get it out. Learn what other nations are doing to power their countries and use their unique resources. America, after all, is a melting pot of climates, terrains, and geological phenomena, not just people.

INTERMEDIATE:

- Take the next step. What's applicable? Practical? Doable within the context of your local political system?

ADVANCED:

- Whether you're an entrepreneur, a politician, a student, a community leader, or just a concerned citizen, dare to dream. Then formulate an action plan to make that dream a reality.

6 THINKING GREEN

Keeping It Simple, Upping the Value Equation, Energy Coaches, Starting Them Young, and Learning from Apple and Coca-Cola

In Portland's laid-back, unpretentious, beautiful neighborhoods like Laurelhurst, plums, peaches, and berries practically drop into your hands from the trees and vines overhanging the sidewalks. Food carts fill up old municipal parking lots. Indie coffee shops crowd seemingly every corner, and the beer brewed there stands out from almost any microbrew around. There are few other places better for a book lover to pass an afternoon than Powell's. Denver's Tattered Cover *maybe*—but Powell's ain't bad.

Seattle has more scenery, better PR, and Puget Sound lies right at its front door. But you don't have to drive long out of smaller, calmer Portland to get to Mt. Hood to the east or Astoria to the west, where Lewis and Clark finally fetched up just over 200 years ago. If it's water you like, the confluence of the Willamette and Columbia Rivers is no small change.

Portland, someone once wrote, is where young people go to retire. A good line, but wrong. Portland is where young people go to enjoy some of the perks of the Golden Years—a leisurely start to the day, easy ambition—while still engaging in meaningful, sustainable commerce. In Portland, people have their morning coffee and maybe a scone while tapping away on their laptops. Then they climb on their bikes and pedal off to work or hop on the light rail for a less-effortful ride downtown.

Portland comes about as close to heaven as a committed crunchy is likely to find. (Plastic bags? The horror! They've been banned from the city's major food stores.) Even hippie dropouts in this alternative Nirvana have a decidedly eco-caste here. Not long ago, a friend sent a photo of a decommissioned school bus turned into a rolling living quarters by some of the city's more marginal residents. Inside sat two sofas that you never wanted your mother to see. A crossed pair of skis served as a kind of ship's figurehead, but this was a land craft all the way. In fact, the old yellow of the school bus had been covered almost entirely in artificial turf. (Talk about green.)

Sneaker makers Adidas and Nike might be the businesses most associated with Portland, but this City of Roses is also home to the US headquarters of Spanish-based Vestas Wind Systems, one of the world's fastest-growing clean-energy companies; the highest concentration of solar installations per capita in any comparably sized city in America; the nation's highest percentage of commuter bikers; and, not coincidentally, the very forward-thinking Energy Trust of Oregon.

The Trust's Betsy Kaufman—a bona fide hero of the movement, living on the frontlines of a truly forward position in the clean-energy future—first called me back in 2005. She had a problem.

"We have a tremendous appetite for solar here," she explained. "People just love it. Every month, the Trust does a seminar to teach people how to do solar. We consistently get one hundred people—a huge crowd, and everyone is enthusiastic."

"That's great," I said. Solar seminars are lucky to draw in the mid-teens, but Betsy wasn't quite through.

"In the end, almost nobody signs up for solar. This doesn't make any sense."

She was right. It made no sense whatsoever, so a few weeks later, Lyn Rosoff, SmartPower's marketing director, and I—within sniffing distance of fabulous falafels and deep-fried anchovies—walked into a drab Portland community center that smelled like stale coffee for the Energy Trust's latest solar seminar.

Betsy was right—everyone was enthusiastic. At the start of the seminar, if they had been any more jazzed about photovoltaic cells and solar hot-water heaters, they might have jumped right out of their skins. She was right about one other thing, too. Four hours later, when the seminar finally ended, not one of the hundred-plus attendants indicated for sure that he or she was about to buy a solar installation of any kind.

"Why?" Betsy asked.

The reason was pretty simple. The first two hours of the four-hour discussion about solar weren't about solar. They covered energy efficiency: how you need a good roof, why you should install good insulation, the benefits of weather stripping doors, changing the light bulbs, attacking that phantom load, etc., and what the likely price tag of all this upgraded energy efficiency was going to be. (Not cheap.) After a break, an hour covered what people had actually come for: solar—photovoltaics and solar water options, what they could do with different houses and roof spaces, and what they could expect to save. Then there was another break. In the last twenty minutes, someone finally mentioned the unmentionable, the actual *actual* reason that most of the group was there: the cost of going solar. (Gasp!)

People were showing up who wanted to buy solar. They might be willing to get rid of their second refrigerator eventually or blow in extra insulation, but right now they were focused like laser on solar. The choir was already in its stall.

"Yeah, but it doesn't make sense to do solar if you don't do efficiency first," Betsy said. "Otherwise, the clean-energy heat will fly right out the door just like the old dirty-energy heat used to."

Imagine a car dealership that spent the first half of a sales pitch talking about how much gas was going to cost you, what it would run you to rotate the tires and change the timing belt, all that stuff. Then they tried to sell you the car. You'd never buy it because they had already convinced you that you couldn't afford it. Same thing with solar, I told Betsy.

People *want* solar, but the people who want them to have solar too often get in the way. No more, at least in Portland.

Those old four-hour efficiency-and-solar gatherings have streamlined down into two hours focused on solar and nothing more. You walk in, and the facilitators say, "We're not going to do Solar 101. You guys are motivated, you know the facts. Instead, we're going to have a discussion, first, about the incentives in the state and how much it's going to cost, and then we're going to break it down to how much it's going to cost you every month."

Why every month? Because to go back to cars again, when you walk into a dealership, the salesperson doesn't say, "I'm going to get you in this baby for a mere $45K!" No, it's, "This great car can be yours for only $200 a month!"

Solar is no different. If you hear that it costs $150–200 a month, you think, *That's cool. I can handle that.* You look at your electric bill and think, *OK, almost a wash.* You'll still get an electric bill. It won't be zero, but it will be significantly reduced. Look at it the wrong way, and solar can be as daunting as buying a Porsche on a schoolteacher's starting salary. Look at it the right way, and solar often just makes sense.

Define, Delete

When we asked 160 respondents who had previously expressed an interest in solar why they hadn't yet installed it, only 1 percent said they had lost interest, could achieve desired savings through efficiency alone, or weren't certain the technology was reliable. Three percent had trouble finding an installer, four percent said the process was too complicated or they couldn't achieve the savings they had wanted. Seven percent learned, after researching the matter, that their home or roof wasn't right for an installation. The largest group by far—47 percent, almost half of the respondents—said that initial out-of-pocket costs were just too high.

Now, keep in mind that our Oregon polling didn't attempt to define the price point at which solar becomes a pipe dream, but my decade of experience puts it at just about $20,000. At $19,999.95, many homeowners can still convince themselves that they can swallow the cost. Kick the price up into even the lowest twenties, though, and the mind reaches for the delete button.

Auto dealers get around this same stumbling block by talking relentlessly about monthly payments. Make sticker price seem like a trivial detail. Numerous solar "counseling services" like the Energy Trust of Oregon have adopted the same approach. Now, Internet solar-cost calculators need to get the religion as well.

One popular online calculator (that shall remain nameless) asks for your zip code, power company, either your kilowatt-hour usage per month or your average monthly bill amount, and the percentage of that you want provided by solar. Then, it chews that up and spits back a mini-chart that shows "solar radiance," whatever that is (the industry still has a propensity to use terms normal people don't understand), the size system needed to achieve your goals, and the necessary roof space, followed by the estimated total cost (almost always stupefying) and the "post-incentive cost," still stupefying.

What that website should do is break that final figure down into digestible payments. A $25,000 loan, for example, payable at 4 percent over ten years will run you roughly $245 per month. The calculator should go further and automatically deduct from that number the projected monthly savings on your electric bill. The system still isn't free, but now you have a figure that doesn't make your hair stand on end.

Government, utilities, and private enterprise also need to do more to attack, in very specific ways, the price-point barrier. Some of all three have launched extraordinarily creative programs. California, for example, offers multiple instances of solar approaches that deserve emulation nationwide. In some cases, they're only pilot programs, but they show us what can happen when renewable advocates bring to their cause the same mind-set and mental energy that goes into selling soda, shampoo, and pet rocks.

Berkeley launched a pilot program that allowed homeowners to deduct the costs of solar installations from their property taxes. Was it a success? Try nine minutes. That's how long it took to run out of available funding. Dime bags of weed don't sell that fast in Berkeley—not even back in the days of Turn On, Tune In, and Drop Out.

The creative (but unfortunately named) SMUD—Sacramento Municipal Utility District—offers another laudatory program that dematerializes solar ownership altogether. Through its SolarShares program, SMUD effectively makes solar power available to owners or renters without adding a single installation to a roof or anywhere else. In the spirit of the locavore movement, SMUD generates the power on a local solar farm, sends it out through the grid, lets customers pay a fixed monthly price for solar based on their electricity usage, then credits that amount against their bill for non-solar power.

That's a somewhat complicated explanation, but SMUD has an online solar calculator (and a separate, nifty clean-power estimator) that makes this easy. Step 1: Enter your most recent SMUD bill or your best estimate of the number of kilowatt-hours (kWh) you use monthly or annually. Plug in $250 a month, and the calculator will tell you that (a) you use about 30,000 kWh annually, (b) have three SolarShare options, a virtual 2-, 3-, or 4-kilowatt (kW) system, and (c) exactly what each option provides and costs. The 2-kW system, for example, would replace approximately 12 percent of your non-renewable electricity with solar power, for a monthly cost of $66. But once you deduct the $53.14 energy credit for going solar, you would have a net addition to your monthly SMUD bill of $12.86. The larger 4-kW system would replace about a quarter of your electricity use with solar at a net monthly cost of $25.72. Quick, simple, clean, direct, low numbers, and in the end you know exactly what satisfying your conscience with renewables is going to cost.

Another California-based outfit, SolarCity in San Mateo, takes the sting out of purchasing a $30,000 roof-top installation the same way Toyota takes the sting out of purchasing a $30,000 Avalon by letting you lease the system instead. But in key ways, SolarCity is offering a far better deal than any auto lessor. Your monthly solar-lease bill, after all, is paying for a roof system that is feeding electricity into your house, which in turn is reducing your monthly utility bill, perhaps by more than the lease payment itself.

SolarCity claims that a 4-kW system installed on a typical three-bedroom house is likely to reduce utility bills from $200 to $60 a month. The cost of the lease is $0 down and $110 a month. Net savings, $30 a month. Not only is that a deal, but, because the cost is locked in for the length of the lease, net savings increase every time utility rates rise, which they've been doing at an average rate of about 5 percent annually. To sweeten the offer even further, SolarCity's SolarLease monitors the output of each system that it installs, alerts lease holders if the system appears to be underperforming, and performs necessary maintenance, gratis. Leases are renewable in five-year increments, but, if you move or don't like the system, the company will remove it at no charge once the lease expires. How's that for a deal?

Over the typical thirty-year life of a solar-roof installation, SolarCity does just fine. Paying $110 per month works for you in the short run, but stretching those $110 a month payments over the life of the installation gives SolarCity $40,000. In the end, you own nothing—just as you don't own a single bolt of a leased car once you turn it back in—but on the other hand, without breaking the bank at the front end, you're likely to have three decades of trouble-free, clean, renewable green energy running through your home wires. Not a bad reward.

SunEdison, based in Belmont, California, offers not one but three lease options on its residential solar insulations; nothing down, your choice down (to lower the monthly payment), and 100 percent of the lease down, which locks in cost, lets someone else worry about equipment and maintenance, and, like all the other leasing arrangements, adds increased resale value to the other savings that going solar brings. SunEd's website references a study by the National Appraisal Institute that found home resale values rise $20,000 for every $1,000 reduction in total energy costs.

With SunEdison, you're also getting an industry leader in every phase of solar—home, corporate, government, everything. As of spring 2012, the company's website counter showed that total SunEdison installations of all sorts had generated an aggregate of over 800-million kilowatt-hours, equal to the energy output of 1.1-million barrels of oil or 325,000 tons of coal, while avoiding 1 trillion pounds of CO_2 emissions, equivalent to planting eighty million trees or *not* driving 1.3 billion miles at 25 mpg.

The Four P's

Price is just one of the classic "four P's", along with product, place, and promotion, that affect where, how, and why we buy things. With renewables, product is taking care of itself. Today, photovoltaic films that cling to your roof instead of sitting on top of it; tomorrow, fuel cells like those powering NASA space capsules. No one can say where it will end, but the genie has definitely left the bottle.

Price, as we've seen, is also in steep decline, and programs like SMUD's virtual-roof installations and SolarCity's and SunEdison's leased ones are bringing renewables ever closer to even our more modest household budgets. As the cost of generating power from traditional fuels rises, the cost gap between fossil and renewables narrows. Here, too, the future is on the side of green energy.

Place and promotion are different matters, but the two go hand in hand. Renewable delivery systems—photovoltaic panels, wind turbines—don't sit on the shelves or clog the aisles of your local big-box home-supply store. They're not browsable in the old-fashioned sense. For us to find them, place and promotion have to work together.

Arizona Public Service—Don Brandt's admirable utility, which we saw in Chapter 4—reduced demand by encouraging its customers to switch to solar-powered hot water. Hot-water heaters are the second biggest electricity hog in the average house, after refrigerators, so increasing the number of conversions means that APS has to build fewer new plants, raise fewer new towers, and string fewer new miles of utility lines to a population exploding at the seams. Solar hot-water heaters also make sense for consumers. At an installation cost as low as $5,000, they can pay for themselves in as few as five years.

How can you lose with a deal like that? You'd think you couldn't, right? But that's exactly what was happening. APS was spending thousands of dollars advertising the idea on regional television, and solar hot-water heaters were flying off the shelves like lead pigs.

Once again, the reason was obvious. No one sits in front of the television, watches an ad, then turns to his wife and says, "Honey, forget that new diamond pendant and our anniversary trip to Vegas. We're buying a new solar hot-water heater!"

Hot-water heaters are a crisis purchase. The basement floods, or the utility closet is soaked. There's no chance of taking a shower or washing clothes or dishes until the problem's fixed. The plumber shows up, takes a look, scratches his head, and says, "Your hot-water heater is a goner. You want the same model? Something a little bigger?"

That's the purchase moment. APS can advertise until they're blue in the face, but that single moment when the plumber lists your choices and the rug beneath your feet is soaking wet is when you're going to make your decision. So the next time a plumber says to you, "You want the same model? Something bigger?" ask "Do you have something solar?" (If APS is smart, they'll reward the plumber, the same way you'd reward any other kind of commercial referral.)

Upping the Value Equation

All four P factors—price, product, place, promotion—create a value equation, the relationship between cost and your expectation of the benefits a product might convey. If price alone carried the day, we'd be knee-deep in products made only in low-wage markets like China, but purchase decisions are more complicated than price. We also want to know what a product will *do* for us. Will it make us feel smarter, cooler, chicer, sharper? Or, alternatively, will the neighbors laugh at me, will my spouse think I'm a sap, my kids think I'm a nerd? This is the Coca-Cola test I mentioned at the beginning of the book. The stuff we buy isn't just stuff. It's a form of conversation, a way of representing ourselves, and speaking to those around us.

Our Oregon polling found that among those who have purchased solar, only 14 percent identified themselves as the first to take risks on new technologies and innovative methods, while 41 percent said they try or buy new technologies only after seeing that they work. In other words, you don't have to be a gearhead to get into solar, and most people getting into solar aren't gearheads.

There are many reasons to do it, and remember that they're not mutually exclusive. One survey we did compared regional differences between Arizona and Oregon in motivating consumers to go green. In Arizona, only 3 percent cited global warming compared to 13 percent in Oregon. Wanting to reduce fossil fuels was roughly equal in both markets, but being good for the environment won by more than double in Oregon—36 percent to Arizona's 17 percent. But on the financial side, 17 percent of Grand Canyon Staters cited saving money vs. 12 percent of Oregonians. Energy costs over time, though, reversed the environmental landslide—31 percent of Arizonians cited this as opposed to only 13 percent of Oregonians.

A multitude of factors might explain the differences—age, crunchiness, etc.—but the takeaway is obvious. It doesn't matter why you go clean, as long as you go clean.

Send Me In, Coach

You know the storyline: The kids can't throw the ball, catch a pass, or run five feet without tripping over their own feet. Then the coach appears—heart of gold beneath a gruff exterior—and everything changes. The losing attitude slowly evaporates. The kids learn that if they play together as a team, rather than as individuals, they might win. Then comes the big game against a team of giants walking the earth. The kids tremble as they watch the other side warm up.

In the locker room, the coach delivers his pre-game talk. "They put on their pants one leg at a time, just like you." "There's no 'I' in 'team.'" Platitudes mount, but deep into the fourth quarter all looks lost. Then coach peers down the bench, spots the kid who never does anything right, calls him to his side, bucks him up, and sends him in to take the final snaps, and the little runt is a natural! He fakes a handoff, races down the field for twenty-five yards, and passes for another first down. Now, it's the final play, trailing by five. He adjusts his glasses, takes the hike, fades back, and you know the rest. Not only is the team transformed, so are the parents, school, town, county, and state. Losers are winners who haven't yet won (or something like that). Only in Hollywood, right?

Not really. This is what SmartPower has been doing all over the country with our energy coaches by upping the value equation and turning losing propositions into winning ones. Here's how it works:

You want to go solar and increase your energy efficiency, but you don't know how or even if you can handle all the variables. We send you a personal energy coach who comes to your house, sits you down, and says, "Look, you have twenty things to do to meet your goals, and I'm going to stay with you for the next three years until the job is done. You don't have to pay me, and a window installer isn't paying me either. I'm here for one reason: to help you get where you want to be."

He begins by categorizing the work that needs doing, easy stuff, not-so-easy, and hard. But he doesn't stop with a simple list. Take the second refrigerator. We make getting rid of it sound easy. *Poof!* It's gone. But dealing with that second fridge is hard for a lot of people. You've got a lot of stuff in there. If it's in the basement, who's going to haul it up the stairs? Can you recycle it somehow? If not, how are you going to get it to the dump? The energy coach understands your dilemmas and has answers.

"The utility will come and take it away for you," he says, "and they'll give you $40 for it." Many utilities do this, although few advertise the service.

"Great!" you say, but what you're really thinking is (a) *I'm never going to call the utility because a human being never answers the phone*, and (b) *How can I explain what I want to a robot?*

Here again, though, the energy coach knows your fears. He's done this before. "I'm calling them right now," he says, and a minute later, "They'll be here Saturday for the pickup."

If the *only* thing you do is get rid of that second refrigerator, you're going to save $20 per month, $240 per year, and so on—not to mention all the CO_2 emitted to produce the energy that you're not using anymore.

That's just the beginning of what the energy coach is going to do for you. He'll tell you about the tax incentives, every little deduction out there. He'll understand your priorities, which might well place the non-energy efficient outlay of painting the house ahead of the energy-efficient expenditure of dense-packing the walls with insulation. It is, after all, your life, your house, your money. His job, like any coach, is to listen and adjust, not dictate.

Trouble is, if SmartPower hired energy coaches for everyone in America who needs one, the company and anyone else remotely connected to this hare-brained scheme would go broke before sundown. But here's the really good news, what brings all this down to scale: energy coaches are contagious.

People who have home-energy audits become very motivated, just like people who buy solar. They want to brag. They grab their neighbors and say, "Look, I got rid of this fridge, I put in a couple of stupid lightbulbs, and this is how much money I'm saving!" You've now become an energy coach, or as we say, a clean-energy ambassador. You're telling your neighbors that the utility gave you forty bucks for your beat-up fridge.

"What?" your neighbor says. "I didn't know that!"

Then you lead your neighbor down the hall to the utility closet, open the door, flip a switch to illuminate the inverter, an incredible little box where you can actually watch the sun's rays from your roof translating into electricity for your appliances. And that does the deal, just the way door-to-door vacuum cleaner salesmen used to close a deal in the 1950s by putting a clean white handkerchief over a vacuum nozzle and running it over a sofa or rug, then showing the black circle where the vacuum had sucked out a world of dirt from what was supposedly a well-kept living room.

But the revelation doesn't stop with the inverter or your neighbor. He passes this amazing news on to his neighbor, and so on. Before you know it, that modest investment in a real energy coach has produced an army of ambassadors, and going green and clean has gone viral. Ambassadors become emissaries to foreign neighborhoods; emissaries become ambassadors to the next town over; and emissaries go statewide. Community by community, we *create* a tipping point.

In Arizona, where we field tested the solar coach concept, we didn't try to tip the whole state in one year, just as we didn't in our initial foray into Connecticut. Instead, we went to individual communities and said, "We want 5 percent of this community to go solar by 2015. It's 2012, and already five communities have done it." Then we went to the newly empowered solar owners and said, "Hey, can we invite ten people over to your house? We'll pay for the coffee and donuts." They said, "Sure I love my solar." Next thing you know, Happy Homeowner is showing her neighbors and friends these really cool money-saving devices on her roof and wall. She's bragging about her free hot water, and neighbors and friends are thinking *This is really interesting*. "Who did this?" they ask. "How did you get this?" And our ambassador-without-portfolio says, "Use my contractor, he's right here."

That's how you sow the seeds of future success—just the way another company did almost from day one. You might have heard of them: Apple, Inc.

Bend It Like Apple

The tributes and commentary that poured in after the death of Steve Jobs focused on his ability always to stay one step ahead of the market; to create the market for everything from micro-music gizmos to pad-like PDAs and then let the market chase him. He was a marketing genius in the same way that Einstein was a genius of relativity—not absolutely right, but absolutely fearless and pretty damn good. But truth told, the real genius of Apple was embedding itself a generation ago in classrooms.

If you are young enough, this will come as a shock, but Apple wasn't always the coolest thing on earth. Way back when, the company had a tiny share of the personal computer market and a business plan that all but guaranteed that Apple would flourish only among creatives and techies, living its entire lifespan on a miniscule little Silicon Valley island far from the raging ports of commerce. Except for one thing: Apple offered teachers and educational institutions deep discounts, which landed boxy Macintoshes and, later, sleeker but still huge iMacs in classrooms, many in upscale school districts, where impressionable children whose families could afford home computers were forming their first real attachments to computer hardware. The rest, as they say, is history.

That's what we're trying to do: hook people on clean energy and renewables while young and impressionable in the hope, and, really, sure knowledge, that this attachment will yield a lifetime of reward.

With the generous help of friends in the renewable industry, we've installed solar at my children's elementary school just as American University has done on so many of its flat-top buildings, with a simple rooftop array of photovoltaic panels. Those panels won't come even close to meeting the energy needs of the entire school, or really much more than one classroom, but the installation comes with a free website that shows, 24/7, how much power is being produced. The teachers can call it up on their computers and use it as a teaching tool, or the students can check it out at home, probably when they should be doing their English homework. However they do it, everyone in that school is receiving a basic message: that bright, hot thing in the sky during the day isn't just the sun, it's a power plant that can light our buildings and recharge our laptops without exhausting the earth's finite resources.

My kids' elementary school represents one example of a much larger green-school movement that's helping to change behaviors at virtually no cost, and with huge savings for school systems and the taxpayers who support them. One article in the *New York Times* in the summer of 2011 estimated that the simple act of putting up sticky notes around the school that reminded kids to turn off lights when they left a room cumulatively was saving tens of *millions* of dollars nationwide. Just as important, kids take that behavior home and turn it into lifetime habits.

I spent most of my teens and twenties as a helpless energy slacker, probably rebelling against my parents' energy stinginess. As the pendulum swings back the other direction, my kids are becoming more like their grandparents. Even though my wife and I try hard to set a good example, it's often not enough.

"Daddy, you left the kitchen light on!" I often hear, and it's like déjà vu all over again, except with a high little voice instead of my dad's deep growl.

Young children are malleable, formable, open to suggestion. Teenagers are something else entirely. They're definitely a separate species, but they are trainable—assuming you didn't sell them as suggested at the end of Chapter 2. We paid high schoolers across the country to take part in a research project. We want you to try to do one of three things every day to conserve energy, we told them: Shut down your computer, TV, and cell-phone charger when not in use, take a shorter shower, or idle the car less. Then we want you to go online (as if it takes any effort to get a teen online) and tell us what you did.

We gathered more than a thousand diary entries. Naturally, teens being teens, they didn't take all our suggestions. Cutting the phantom load from computers, TVs, and cell-phone chargers was their number-one choice. Shorter showers—painfully close to hardship duty for most of them—slipped to number three, while the nebulous "other" came in second. But the point is, just about all the high schoolers we surveyed did *something* proactive to reduce their energy usage and most of them blogged about it afterward. They, too, became cyber ambassadors and social-media agents for clean energy and renewables.

For all of their sullen moodiness, teenagers want to be inspired. They want to feel smart about the clothes they wear, the music they play, the websites they visit, and the energy they use. Imagine if we could harvest all their energy smartness and send it out via Facebook, Twitter, YouTube, and all the rest of the sites they visit. That would be an absolute game-changer.

As you saw in Chapter 2, SmartPower launched a program that engaged more than 20,000 students at 465 colleges and universities in reducing their collective carbon footprint by 19 million pounds and creating $4.52 million in energy cost savings. The program, America's Greenest Campus, used a sophisticated web platform to provide personal energy-saving recommendations and tracked both individual and collective progress, piggybacking on viral and social marketing and endorsements by hip-hop mogul Russell Simmons and then-Internet phenom Obama Girl. (Remember, this was 2009.) To the school with the most carbon reductions and the school with the most participants went awards of $5,000 each. But you can't sell air conditioners to Eskimos who don't want them. What America's Greenest Campus really did was corral and ride the natural desire of students to reduce dependency on fossil fuels, cut down their own admittedly wasteful power consumption, and give energy efficiency a fighting chance.

Barbarian at the Gate?

This hard, *Mad Men* emphasis on pushing solar, wind, and geothermal, as well as energy efficiency to reduce demand, can get a little grating, especially to those for whom going green is virtually a holy cause. Efficiency and renewables ought to sell themselves, right? No, wrong. In fact, sheer nonsense.

Robert Woodruff, who became CEO of the Coca-Cola Company in 1923 at the tender age of thirty-three, didn't elevate his brown fizzy drink to the world's most-celebrated brand by sitting in his Atlanta office and hoping people everywhere would discover a product nobody really needed. He wasn't even a promising choice to take over the company and probably wouldn't have gotten the offer if his father hadn't put together a syndicate to buy the company. But Woodruff and the team he assembled understood the marketplace as maybe no one before or since has.

Santa's red jacket? It's Coke-red in case you haven't noticed. Santa frequently dressed in green before Coca-Cola let loose a classic series of Christmas ads that linked the beverage and the world's favorite gift giver. In 1928, Woodruff sent a thousand cases of Coke along with the US team to the Olympic Games in Amsterdam. A Coke a day keeps the gold in play. In the aftermath of Pearl Harbor, Woodruff committed his company to providing "every man in uniform [with] a bottle of Coca-Cola for five cents, wherever he is and whatever it costs our company." Patriotism? Maybe. But as Coca-Cola followed servicemen across the world its global presence grew, too.

Energy Secretary Steven Chu loves to cite the example of granite kitchen countertops. One day they didn't exist, the next, Formica was more passé than Nehru jackets, and you couldn't sell a new house anywhere in suburbia without a sea of granite in the kitchen. That's what needs to happen with renewables. They have to become the energy equivalent of whatever is going to become the next granite countertop. Not an oughta-have, but a gotta-have.

WHAT YOU CAN DO

BASIC:

- Start doing the simple, easy things that make you energy smart: Turn off your computer at night, unplug your cell phone chargers when they aren't charging anything, and completely power down the flat-screen TV when you go on vacation or even away for the weekend.
- Look around for a solar home tour—the gadgets alone are worth your time.
- Get your kids involved in the clean energy conversation, too, if they haven't already involved you.

INTERMEDIATE:

- Get involved. Attend a local renewables workshop or seminar.
- Be greedy. If it's time to sell your house, remember that every $1,000 of solar improvements add an estimated $20,000 to your home's value.

ADVANCED:

- Move to Portland, Oregon.

7 THE HARE AND THE TORTOISE

Home Runs, Singles, and Clean Energy by the Numbers

As a child, I was quite close to the cutting edge of the NASA space program. Literally. Judi Getch Brodman, part of a team of General Electric geniuses who developed the hydrogen fuel cells that helped power the Gemini and later Apollo flights, lived next door to my family's house growing up in Needham. In fact, she's still in the house even though my parents moved on some years ago.

Judi was a mathematician by training, but none of us neighborhood boys held that against her because (a) she was a complete looker—in-house GE publications at the time always referred to her as "lovely" or "perky", and (b) we knew that what she was doing was really, really cool, even if none of us could say what a fuel cell was or did.

In a way, all that's still the case today. Judi is as lovely in retirement as she was when she put in long days with

GE's Direct Energy Conversion Operation. Fuel cells are still mind-blowingly cool, a challenge to describe, and as out there on the edge for clean energy as they were a half century ago when they provided water and other essentials for astronauts Frank Borman and James Lovell during their record-breaking fourteen-day flight aboard *Gemini VII.*

Reduced to the simplest terms, fuel cells create energy out of nothing. In their standard form, they're basically very thin, very light proton-exchange membranes (PEMs) treated with a platinum catalyst on both sides. Supplying pure hydrogen to the PEMs at an elevated (but not extreme) temperature sets off an electrochemical reaction between the hydrogen and oxygen that converts chemical energy into electrical energy. Cathodes and anodes get involved, just as in standard batteries, and the electricity is captured and converted to power, say, a vehicle or whatever.

Fuel cells have long been the dream of renewable-energy advocates because nothing we know is cleaner or greener than hydrogen. The reality of a fuel-cell-driven future, though, has proved elusive. Honda has had a high-performance, fuel-cell powered sedan on the road since 2008—the zero-emission FCX Clarity. But the pure hydrogen necessary to power the car can't be pulled out of the air as you go, and, while the Clarity has an admirable 240-mile range between fill-ups, there's no infrastructure of hydrogen stations in the country outside of Southern California, yet. Price is an issue, too, in part because of extremely limited production but also because the platinum that speeds the electrochemical reaction is itself so expensive. A three-year lease on an FCX Clarity—the only way right now to take possession of one—runs $600 a month. All of which means that Hollywood stars and their producers, directors, and agents are driving most of the fifty or so Claritys on the road today.

The times could be a'changing for fuel cells and their uses, though. Applications on file with the US Patent and Trademark Office show that Apple has been working on hydrogen-fuel-cell technology that would power computers and phones for weeks without recharging. The cells would be considerably lighter and less bulky than the current generation of lithium-based batteries, another way that Apple can make its thin and hip products thinner and hipper still.

The Department of Defense is in on the chase, too. A 2011 report commissioned by the Pentagon concluded that the military "should more proactively evaluate and acquire fuel cell systems." To that end, Secretary of the Navy Ray Mabus spent part of Pearl Harbor Day 2011 on Hawaii's Oahu Island, being briefed on five GM-manufactured fuel-cell vehicles undergoing testing at the Marine Corps base there. Other possible military uses for fuel-cell power include unmanned undersea vehicles and the next generation of aircraft drones. Again, their range would extend much farther than with conventional fuels and their weight would come in at a fraction of what's possible with current technology. Just as important, fuel-cell technology would bring the various service branches in line with DOD mandates to seek out and deploy clean and reliable energy sources. As it sifts through every military branch, this top-down push could radically change the way we wage war and how we deploy our troops overseas.

The biggest holdback to fuel-cell technology has been the cost of producing hydrogen and providing an infrastructure for distributing it, but that, too, could be changing soon. MIT professor Daniel Nocera and a team of colleagues have developed an artificial leaf made of silicon that they say is capable of using sunlight to split water into oxygen and hydrogen, which can then power fuel cells. Instead of relying on expensive platinum as the catalyst, Nocera's leaf, which looks like a thin black magnet, uses a cobalt-and-phosphate compound on one side to pull oxygen from water, and a nickel-based alloy on the other to release hydrogen.

"You drop it [the leaf] in a glass of water and you walk outside and hold it in the sun, and you'll start to see bubbles of hydrogen and oxygen," Nocera told a *Boston Globe* reporter. (Watch Nocera's explication of his team's discovery at YouTube.com/watch?v=Tpot_MSe2g8.)

Marketable applications are still a long way off, but venture capitalists are watching and circling, and with good cause. The "leaves" are made of readily available, abundant materials and require no power other than sunlight to separate water into its component parts. Just as real leaves store the sun's energy so they can get through the night, so these leaves store energy that can be released as needed. Nocera envisions a home solar system that uses photovoltaic cells to power the house during the day and to help separate water into hydrogen and oxygen, then release the hydrogen into fuel cells at night to provide hydrosolar power around the clock, including those dark hours after midnight when you might want to charge an electric vehicle.

Missing Links

Judi Getch Brodman's fuel cell, Honda's FCX Clarity, the Apple and US Navy experiments, and Daniel Nocera's vision of a house that functions as its own solar power station are all incredibly inspiring. But they also remind us of the missing links in achieving an all-renewable, zero-emission future: storage, transmission, and distribution. Green is good—but greener is better. So the question is, how do we get there without waiting for the future to arrive?

Bill Gates likes to say that all the batteries in the world can store only as much energy as the world consumes in ten minutes. That bodes poorly for storage as the ultimate answer, even if a new generation of storage batteries grows to be (as some are proposing) almost as big as the power plants they might be replacing or supplementing. Then again, with better transmission of renewable energy, batteries the size of small cities might not be necessary.

Martin Hoffert, the NYU physicist who forecasted the end of fossil fuel production by the end of this century and who called for a three-prong energy Apollo Project, suggests that we reimagine the whole concept of a national grid. Today, the grid links and mixes centralized American power sources—fossil fuels and renewables—but what if there were an international grid instead, one that used above-ground and sub-ocean transmission lines to connect wind farms in Denmark, the Caucasus, the Andes, and Australia; vast solar installations in the Sahara, far eastern China, and the Sonora desert of the American Southwest; and geothermal sources in Iceland, the Philippines, New Zealand, and elsewhere?

Then renewables truly would be global, and it wouldn't matter at all when the sun went down or where the wind was howling at any one moment or any one place.

Or maybe the future is no grid. Or every person is his or her own grid, or every window its own power plant and grid. That's what Justin Hall-Tipping put forth when he addressed a TED gathering in Edinburgh in July 2011. Hall-Tipping is an investor, not a scientist. Back at the turn of the millennium, he launched Nanoholdings LLC to take advantage of breakthrough energy findings at the molecular level, so he has self-interest in the game. But TED's mandate is to spread ideas, and Hall-Tipping has compelling ones, plus corroborating evidence, to throw around.

Take windows. Light and heat come in, light and heat go out. That's obvious, but what if we could control that equation at will; flick a switch to bring the heat in without affecting illumination and flick it again to keep the heat out, also without affecting illumination? Like other energy entrepreneurs, Hall-Tipping has a solution: carbon blasted with a vapor, then condensed into wire 100,000 times smaller than the width of a human hair, a thousand times more conductive than copper, and transparent, not black as we normally think of carbon. Combine this vaporized carbon wire with a polymer, affix it to a window, flip a 2-volt switch, and the window can keep all the heat out, let it all in, or achieve any point on the scale between, he told the TED audience.

But even that only touches the tip of what an everyday window might accomplish. Hall-Tipping foresees (and has university scientists working on) using more and different nanomaterials to capture the full spectrum of infrared light available at night, convert it to electrons storable *on* your window, thus making it (or any other surface) a personal power plant that holds energy until you need it or converts it back to light and beams it by line of sight to any place you want.

"For that," Hall-Tipping told the rapt audience, "I do not need an electric grid between us . . . The grid of tomorrow is no grid, and clean, efficient energy will one day be free."

Will it? It's hard to say. These silver-bullet, clean-energy-of-tomorrow speculations are inspiring, but that's not how we're going to win this game. We need people thinking and acting at the edge of the possible, maybe now more than ever. But this contest is a marathon, not a sprint; a tortoise's race, not a hare's. The victor won't be a lone genius—or even team of geniuses—ginning up some solution that we lesser mortals can't even imagine. The victor must be all of us, using our wallets and consumer power to bring the freshest, best, most applicable and desirable clean-energy ideas to the marketplace.

At some level, we all know that, and as interim proof, here's a little numerical comparison. Marty Hoffert's spine-tingling video tour through coming energy shortfalls and the need for a new energy Apollo Project had been on YouTube for almost three months when I first watched it. I was the thirteenth viewer. Daniel Nocera did better. His call to fuel cell technology had been on YouTube for four months and had logged almost 11,000 viewers. Justin Hall-Tipping, with the powerful TED brand behind him, far outpaced the other two, logging over 87,000 views in just a few months. But

SmartPower's Obama Girl video, the cute animation that we created to promote the practical suggestions of our America's Greenest Campus campaign, logged almost 500,000 views in a single week.

Lesson 1: Great ideas lead to paradigm leaps, but you need to get the message out.

Lesson 2: Until rubber meets road, we're all just talking through our hats.

Lesson 3: Our Arizona Solar Challenge.

Clean Energy by the Numbers

In 2010, we launched the Arizona Solar Challenge with a simple proposition: we would designate any participating city across the state as an Arizona Solar Community once 5 percent of its owner-occupied homes had installed solar roof panels. Communities had five years to hit the mark. That was it. No cash prizes, no symbolic free solar installations for qualifying communities as we did in Connecticut and are doing in Massachusetts and elsewhere. Just the chance to put a sign up at the entrance to the town, next to the Rotary and Kiwanis signs, declaring that this was an Arizona Solar Community.

We created the infrastructure to make this happen using our COR: C for Community outreach, O for Online platform and social marketing, and R for Rewards and incentives.

We went to mayors and/or city councils and said, Hey, can you commit to 5-percent solar by 2015? We're not asking for money or manpower. You won't have to burn any political capital, but you'll be showing leadership if you do.

Once we had the buy-in at the local government level, we connected with community environmental groups, churches, and service clubs and got them to use their networks to push the goal along.

The O part—online platform and social marketing—is the connectivity that holds this network of interested, motivated parties together and gives it the room and the means to expand. We set up an Arizona Solar Challenge website where community members could tweet with one another about solar, join the challenge's Facebook page, check on the progress of their own communities, or, what most people do online these days, zero in on the vital information (where do I find an installer, etc.) that used to take weeks to find.

Specific rewards in this instance amounted to nothing more than the chance to put up that "Arizona Solar Community" sign at the city line. In a devastated real-estate market, little things like that often mean *a lot*, but incentives are all over the place: the chance to save money; to do something good for the community, the country, and the planet; or just to be cool. Whatever it takes.

Then we orchestrated the ballet I described earlier: We found that alpha couple who first installed solar on their roof. They invited ten of their neighbors over to hear the story. They had their contractor on hand to answer questions (and of course drum up more business). We did all the heavy lifting—provided coffee and donuts, handled the invites, had someone explain financing possibilities, brought in a solar coach who knew everything there is to know about solar power but could talk about it in plain English and *wasn't* selling anything.

But don't take my word for it. Let's look at the numbers. In the two or so years since the Arizona Solar Challenge began:

- Fourteen communities have signed on, and five have already met or passed the 5 percent threshold.
- Sixty-one media stories have been generated.
- One hundred community events have been held.
- We've designated 246 solar ambassadors.
- Fielded 1,306 solar coach requests.
- Logged over 11,000 customer interactions.
- And the central issue: Arizona Solar Challenge has been instrumental for over 4,000 solar installations, nearly 75 percent of all solar installs in the state and about 3 percent of all solar installations nationwide ever, from what is essentially a pilot project.

The numbers, though, tell only part of the story. By increasing demand for solar at the grassroots community level, the Challenge has created jobs for local contractors and installers; provided public education on the benefits of clean, renewable energy; and reduced reliance on other forms of energy that cause pollution and reduce air quality in communities all across the state, whether they took part in the Challenge or not.

Can you do this yourself? Sure. We haven't patented our process. In fact, we *want* it to spread. You can do the same. Set a solar goal for your community, a renewables goal for your school or apartment building, a phantom-download-reduction goal for an office or dormitory. Do the legwork to get buy-in from local leaders, whether they're elected officials, a co-op board, residence deans, or whatever. Work the online platforms, and the offline ones, too. (Cork bulletin boards and under-the-door flyers still have a place in modern communication.) Come up with the sort of rewards that will have meaning in your own setting. And *then* roll up your sleeves and get to work. Hydrogen-fuel cells might someday change the world overnight, but until then, we need to make the world more green one day, one person, one renewable kilowatt at a time. And we can.

Memo to Washington

Paul Tsongas, the late Democratic senator from my home state of Massachusetts, was once my political mentor in Washington. We first met on St. Patrick's Day, 1991, when I was a young aide on Capitol Hill. By that summer, I was working on his presidential campaign, and I would stay with it and him for the next seven years.

In a sense, Paul's presidential campaign never should have had a chance. His campaign was underfunded; he was up against a brilliant vote-getter in Bill Clinton; he brought too readily to mind the spectacularly failed campaign waged by another Bay State Greek, Mike Dukakis; and there was always in the background the question of Paul's health. He had nearly died of cancer five years earlier, and, as it turned out, he succumbed from complications five years hence.

What Paul Tsongas did have going for him, though, in spades, was character, vision, purpose, and hope. He didn't buy into the cynicism that politics is some infernally corrupt game. He believed deep in his heart that politics was the engine that drove democracy—the collective decisions that keep our system alive, vital, and progressive so that it can serve the common good. He also believed that marketing an idea was best done on the grassroots level, person to person, neighbor to neighbor, community to community.

In Paul's spirit, I offer the following memo for the use of any and all readers, regardless of political persuasion, who believe that green is good, its time is now, and clean energy is everyone's problem and everyone's opportunity. You can find a copy of this at www.smartpower.org/memo.

Subject: Green Is Good

Dear [Elected Official]:

Every election cycle we relearn the old truth that all politics is ultimately local. So is all behavior change when you get right down to it.

The fight for clean-energy needs is not about federal legislation or government mandates. It's about changing behaviors so people can help themselves. Government has a long and noble history of doing this. Americans didn't become sensitive to the dangers of forest fires because a law decreed it. Smokey the Bear—that great federal ad campaign—did the job just fine. Similarly, government marketing programs at all levels have turned the tide against cigarette smoking far more than any combination of federal mandates, and state and local laws. No adult is forbidden from buying cigarettes, but education and advertising programs have given us Americans the power of informed choice when it comes to tobacco.

The Civil Rights Act of 1964 was an enormously important piece of legislation, but what changed people's perceptions was the community organizing of Martin Luther King Jr. and others. That's what got people asking: Why can't we all sit at the same lunch counter and use the same water fountain? Those changed perceptions ultimately changed behavior in a way that no law itself could have done.

We can do the same thing with clean energy, and we must. The interest in clean energy doesn't perk up because Big Brother commands us to be interested in clean energy. It perks up because a fellow Lions Club member or book-club buddy or Sunday School teacher is suddenly thinking about putting solar on her house.

I urge you to put the clean-energy push on the ground where it belongs. Help volunteer groups go door to door, neighborhood to neighborhood, community to community so they can show people the hows and whys of clean energy. Give clean energy the chance to sell itself, and you'll be doing more than spreading renewables, you'll be creating jobs for contractors and salesmen, and conditioning the next generation of Americans to clean-energy solutions.

Sincerely yours,

WHAT YOU CAN DO

BASIC:

- Create your own neighborhood clean-energy group.
- Write your local, state, and federal government officials, including the president, our community organizer in chief, and urge them to lead a clean energy grassroots campaign.

INTERMEDIATE:

- Take the time to educate yourself about clean energy and energy efficiency. Instead of surfing TMZ.com for Hollywood gossip, visit websites that will teach you how to be energy smart. Your electric utility's website likely has a ton of programs that will save you money and be environmentally friendly.
- When you do install solar, invite your friends over, and the contractor, etc. Brag a little. Make it a party!

ADVANCED:

- Become a solar ambassador and join our efforts.

8 EVERY PERSON'S GUIDE TO CLEAN ENERGY

The following are some basic questions about, and steps toward, improving household efficiency and using renewables, both of which will help you save the money that you spend by powering or heating and cooling your house. This is by no means an exhaustive guide, and it will probably spur more questions. But taking ownership of your energy decisions is even easier when you understand how much you might spend or save, what small steps you can take to improve efficiency, how different renewables work, and which are right for you.

The Starting Line

Before you walk outside to survey your roofline for a solar installation, gather all of your utility bills for the past year. Figuring out what you want to save means knowing what you spend, and in most cases your money is going to pay both the power company and at least one other utility, maybe natural gas for heat or an oven. If you crunch all of your utility numbers, you can create a total household energy profile, once you get over a little conversion problem.

Your electricity bill displays everything in kWh (kilowatt-hours), which makes it easy to understand how a 2-kW solar array will impact your monthly expenses. But the natural gas that your home uses is measured in therms. Fuel oil is measured reassuringly in gallons, but, although it's easy to visualize your total yearly heating fuel as 500 containers of milk, that image doesn't really help you relate fuel to the kWh you use to power your cell phone.

When you've gathered your annual usage amounts in therms, gallons, BTUs, and whatever else you may have, plug them into OnlineConversion.com/energy.htm, one of several websites that can convert those various quantities into kWh. Those 500 therms of natural gas turn into 14,650 kWh, and the 500 gallons of fuel oil you burned turn into 20,000 kWh. If you want a *really* complete household energy profile, throw in that half cord of juniper wood that you burned in the fireplace, using 9,750 BTUs or 2,857 kWh.

If you want to make your home truly energy neutral, this is essential information. But even if you're just interested in getting a solar array to knock down the electrical bill, you might upsize or downsize a project based on these numbers, or you may realize how much you could save with a specific renewable, like solar thermal. These numbers are also important when considering household efficiency measures. If an enormous amount of your utility expenditures goes toward heating your house, you might look at insulation and air sealing *before* investing in solar thermal. If your house is well sealed, but your central air is burning through electricity in the summer, maybe you need to think about a solar photovoltaic setup or a geothermal cooling system.

It Adds Up

There are plenty of smaller, cheaper jobs around the house that almost anyone can do solo or with one other nonexpert helper. Some of these projects aren't even really work jobs so much as behavior modification. Remember the 10 percent of our electricity bills going to unused phantom load? Now's the time to scour your house for any appliance with a light or clock: computers, DVDs, cell-phone chargers, microwaves, etc. They're all drawing power even when you aren't using them. Unplug devices that don't have long reload times. Rather than unplugging them directly from the wall—a hassle—plug them into a power strip that you can easily switch off. Better yet, in an older house, if you have an outlet controlled by a light switch, plug those components into that outlet. Flip the switch when you leave the room (the hard part), and you're saving easy money on your power bill.

Laptops use less than half the power of desktops, so switching to a laptop can save you up to $35 a year. If you use hot water to wash clothes, switch your settings to warm or, better yet, cold, and save $60 a year. Your laundry gets just as clean without the heat. It takes about one minute to install a low-flow showerhead ($10 and up). In a house where a shower runs an average of twenty minutes a day, you'll save about $80 a year on heating and water costs. That's $10 a month just by installing a new showerhead and washing your clothes in cold water, about two minutes' worth of labor. Pretty quickly these small projects turn into real money in your pocket.

During winter, try gradually lowering the thermostat one degree at a time. A whole day at seventy degrees instead of seventy-one saves 3 percent on your power or gas bill. The average American heating bill is a little over $900 annually, so a 3 percent savings is around $27. If you can handle sixty-eight degrees (just two measly degrees shy of standard room temperature) that's $81 right there. Try turning the thermostat down by ten degrees for those eight hours at night that you're sleeping. Too cold? Put an extra blanket on your bed or try a hot-water bottle—old technology that works great. Flip the heat back up in the morning—or have a programmable thermostat do it for you—and wake up to some coffee and an extra $90.

Ready for more? Let's take a closer look at the kitchen.

Recipes for Savings

Unless you're running a body shop out of your garage or eat out for every meal, chances are that the kitchen uses the most energy in your home. Kitchens today are brimming with appliances—blender, coffeemaker, dishwasher, electric can opener, food processor, microwave, oven, refrigerator, toaster oven, and about twenty more possibilities depending on how or what you cook (bread maker, Crock-Pot, ice cream maker, pressure cooker, rice cooker, slow cooker, and so on) and how often. This puts a bull's-eye on your kitchen as the No. 1 opportunity for energy savings. You can scope out some simple and cheap—$15 or less—ways to save money and energy by asking yourself a few basic questions the next time you're cooking or using other kitchen appliances:

• **Why is my empty stove on?** Preheating is only necessary for foods that rise, like biscuits, bread, and cakes. For almost any other food, it doesn't matter how hot the oven is when you stick it in. Cooking might take a few minutes longer if you start with a cold oven. But when preheating is necessary, it doesn't take the twenty minutes or so that many recipes suggest, particularly if you have gas heat. If you have a display that counts preheating time, you already know this. If you don't, you can buy an oven thermometer ($5) to track it.

If you use your oven's self-cleaning setting, which superheats food particles to ash, consider scrubbing the oven yourself with an oven cleaner ($10) or make one at home with water and baking soda. It's a little bit of work, but remember that the self-cleaning mode is running your oven on maximum for a full three hours. If the thought of making your own cleaner sounds a little too DIY for you, fill a casserole dish with water and boil most of it to steam inside the oven, which also will loosen up that gunk. (You can use the same trick with the microwave or by boiling water in pots and pans on the stovetop.) Depending on how often you use the self-cleaning mode, a little elbow grease can earn you $40 to $100 of savings a year.

• **Why is the oven door open?** Opening the oven door to check your food usually drops the internal temperature by about twenty-five degrees. Sometimes you don't have a choice, but use the light when you can. Also, if you're cooking a roast, keep track of the temperature inside the meat quickly and easily with a meat thermometer ($7 to $15) to avoid taking the roast out too long to test it.

When the door is closed, make sure it's closed tightly to keep the maximum heat inside. Gently use a degreaser ($10) to clean the seal on your door. As with preheating and self-cleaning, your savings will vary widely depending on your cooking habits, but they're there for the taking.

• **When am I cooking?** Any appliance that generates heat or cools you or your food down is going to cost you. Baking during hot summer days creates a perfect storm in which excess oven heat forces your air conditioner and refrigerator to do battle against an enemy within. Cookies in the oven on a hot day can easily raise the temperature by ten degrees, which can increase your cooling costs by 5 percent. If your summer cooling bill is around $300, cooking at night or grilling outdoors will not only feel more comfortable but it will save you $10 to $20.

• **Do I even need to use the oven?** If you already have a microwave, toaster oven, slow cooker, or pressure cooker, those appliances can do an oven's smaller jobs at a fraction of the cost. A toaster oven uses half the energy of an electric oven, saving 50 percent.

• **Why is the refrigerator door open?** Yes, staring into the fridge, frozen in indecision, costs money. Listen to how quickly the compressor kicks on to cool things down again. But a simple, annual task that might not have occurred to you is dusting the fridge. Not the top (although, while you're at it, you might as well) but the back. Pull the fridge away from the wall and clean off those coils. Also, keep all food inside the fridge covered to limit evaporation, which leads to harder-to-cool moist air. Two steps, no money, and your fridge is quieter, healthier, and more efficient.

• **What's in my dishwasher?** Wash only full loads of dishes, and you'll save money on water and energy as well as avoid excess heat production. When it's hot out, run the dishwasher before you go to bed at night instead of during the day. Also, try air-drying your dishes instead of using the heating cycle, and you'll save even more on your electricity bill.

• **Which of my appliances is the biggest money drain?** That would be your clothes dryer. These machines average between 1,800 and 5,000 watts. They draw so much energy that there is no certified Energy Star model because there's no way to make them more efficient. Running one is equivalent to sliding forty cents over to your power utility every time you run a load. Sure, it might be cheaper than a Laundromat (though not that much), but this is your house! If you do four loads a week, like some families, that's over $80 a year.

So how do you save some of that electricity and get back some money? The best option is not using a dryer at all if you can. Line-dry your clothes ($11 and up for a retractable clothesline, $8 and up for twenty-four clothespins) or use a metal drying rack ($20 and up—and better than those cheap wooden ones that wet laundry will sag, warp, and break). Clothes smell better when dried outside, and they last longer when not exposed to the extreme high temperatures of the dryer.

If that's not an option, as with dishes, only do full loads to cut down on the number of times you have to run the machine. Also, unlike with washing machines, where you have to separate whites, lights, colors, and darks to avoid color bleeding, you can dry mixed loads of laundry in the dryer. (Fabric dye doesn't run in hot air.) When you do need to run a clothes dryer during the summer, do it at night.

Tackling the Big Bills

We live all across the mountains and plains and swamps of the country, but no matter your environment, odds are good that you're spending about half of your utility bills on heating or cooling your home. If you live in one of the colder states, heating can account for over 65 percent of your total energy bill, leaving the other 35 percent to your home's lights, flat-screen TV, clothes dryer, fridge, dishwasher, and oven combined. Next time it's hot outside, take a stroll around the house with a few questions in mind:

- **What's that up there?** Is it that century-old relic, a fan? Great! Whether used alone or in tandem with an air conditioner, fans circulate air and keep you more comfortable. A central air system runs between 1,000 and 3000 kWh a month, while a ceiling fan ($100 and up for a quality fan) cools a room using about as much electricity as the lights illuminating it. The breeze created by the fan can make a room feel five degrees cooler. Which means that you can set the thermostat to seventy-four degrees and still sit in a house that feels like sixty-nine degrees. Raising the thermostat that much can save you 20 percent on your cooling bill, creating a pretty quick buyback time on that ceiling fan.

If you're not using A/C, or if nighttime temperatures drop into the seventies, a more comprehensive solution is to fan out the whole house. Opening windows lets in nighttime air. All you have to do is close them up again—including lowering shades if necessary—as soon as temperatures rise in the daytime. The best way to do this is to suck through air with a whole-house fan ($300 and up, installed). It'll exhaust all the hot air through your attic, and moving air feels cooler than still air. A whole-house fan can cut your cooling costs in half, depending on your house's size and location.

• **What else is up there?** Your neck may be getting stiff, but now that we've sorted out your fan situation, check out those light fixtures and the lamps around you. Replacing incandescent bulbs—the ones that mean "bright idea" in a cartoon—with compact fluorescents (CFL) or Light-Emitting Diodes

(LED) (both types of bulbs start at around $10) is a truly great idea. Incandescent lights use far more energy and release nearly all of it as heat—as you and your burning fingers have discovered while swapping out a live bulb. It's virtually impossible to burn yourself on LEDs, which release almost no heat, while CFLs release very little. Both bulbs also use about 75 percent less energy.

At the store, incandescents are much cheaper, but those savings end as soon as you get home. They eat up so much more electricity that the "more expensive" CFLs and LEDs will be saving you money within a year. Swapping out all the bulbs used to light four or five rooms can save you more on your monthly electrical bill than you're using if you regularly run your dishwasher, all while lowering your summer cooling bill. Those LEDs and CFLs also last up to ten years.

- **How did the sun get in here?** Blocking the world's number-one heating source with shades and blinds can make a huge difference in your summer cooling bill, and you can choose from inexpensive roller shades ($20 and up) all the way up to insulated black-out curtains ($50 and up). You can also block sun from the outside by planting deciduous shade trees that lose their leaves in the winter and let the warming sun in. Either way, you'll knock as much as 20 percent off your cooling bill.

Keep It to Yourself

Some of the same methods to cool down the house in the summer work well for heating if you just turn them upside down. You can reverse a ceiling fan's blades to blow down warm air trapped on the ceiling. You can also turn the thermostat down in the winter. In fact, the best way to maximize your savings is by installing a programmable thermostat (starting at $30 and easy to install) so that your heat or air conditioning is working only when you're home or need it, saving up to 20 percent of your heating and cooling expenses.

But staying comfortable inside, whether by heating and cooling, comes down to how efficiently you can control the air in your house. When you want to cool down a house, you turn on a house fan and pull the hot air out through a hole in the ceiling. When you want to heat up a house, you need to keep that same hot air inside, which raises the question:

• **What am I missing?** This is the best question to ask before undertaking any great endeavor, from launching a space shuttle to running errands. It's also particularly appropriate when looking for the insulation that may or may not be tucked invisibly into walls, attics, floors, crawl spaces, and around ductwork. Even if you have a well-insulated house, you can't see it, although you can probably feel it. But if your insulation is in fact just missing, or doesn't have a high-enough R-value—the rating that measures how well the exterior of a house resists the transfer of heat—you've got plenty of energy-saving opportunities and lots of ways to do it.

First, you should check out what the different R-values are going to mean for you. The Department of Energy has a quick Zip Code Insulation Calculator at www.ornl.gov/~roofs/Zip/ZipHome.html that suggests R-values for different areas of your house. Once you've got that settled, you've only just begun. Depending on the space you are insulating, you may need to cut holes in your walls and ceilings, rip out wiring, and take up floors in order to roll out batts of insulation, spray in foam, or blow in cellulose among other possibilities. It's not as easy as popping in a compact fluorescent light bulb. On the other hand, plenty of cost-saving insulation projects start under $100 and are easy enough for the average adult. As you move up in scale, expense and complexity will increase as well. Before you balk at spending thousands on insulation and air sealing, remember: This step could cut your heating and cooling expenses in half, or take 25 percent out of your total utility expenditures.

Tap into Nature

Now that you have a few ideas about how to keep your current house more efficient, you may be ready to think about bigger changes that will allow you to produce your own power, heat, and cooling.

SOLAR

Solar power is probably the most glamorous of the renewables. Harnessing the power of the sun sounds almost mythical. It isn't as grungy as steam or smoke-spewing turbine power, and it receives some of the most generous state and federal incentives. In fact, it begins with a piece of silicon, one of the most abundant elements on earth. In its raw, unrefined form, silicon helps make bricks, gravel, mortar, and porcelain, while a synthesized form, silicone, makes sealants (and formerly breast implants). A highly purified form of silicon goes into the semiconductor technology that makes modern computing possible and Silicon Valley shorthand for innovative technology. Solar-grade silicon isn't as highly refined as what goes into computing technology, but it does serve as a semiconductor, meaning that it conducts electricity when exposed to light or heat—in this case, the sun.

A cell that measures around four or five inches across only creates about .5 volts of usable power, but a solar panel consists of many cells that, when wired together, add up. A panel with thirty-six cells has a maximum power voltage of about 18 volts. The cells, however many there might be, are housed on a panel and covered with anti-reflective film and glass to seal them from the elements.

When you start looking at different panels, some will come in aluminum frames, some in stainless steel, others in plastic. Some will produce a maximum of 200W, others may go up to 250W. Certain panels will be much more efficient—wattage produced per square foot—and will often cost more as well. Many of these differences go back to the variations in materials used to build the cells. Monocrystalline cells, for example, are the most efficient but are very energy intensive and thus expensive to produce. Next down the scale are polycrystalline cells, slightly less efficient but with lower production costs.

There are two other commercially available solar photovoltaic options. Those thin-film or thin-foil panels rolled out onto the roofs of selected Wal-Marts are sheets made out of what are called amorphous silicon cells, a less-than-a-hair's-width-thick layer of silicon sprayed onto various backings, including metal and plastic sheeting. Thin-film sheets are the cheapest and least efficient option and make the most sense when you've got lots of space to cover like a big box retail space. These thin-film technologies also perform better in high temperatures—perfect for hours of baking atop a shadeless strip-mall roof.

At the extreme other end of the range sit bifacial HIT panels—what Sanyo used to encase its office building at the Kasai Energy Park, in Japan. HIT panels use both the most expensive monocrystalline cells on the panel's front and a layer of the thin foil technology on the back of the panel, capturing the sun's light coming (direct) and going (reflected).

Each impressive in its own way, these four different technologies all have to deal with the same limitations beyond the lab. A 200W panel is rated under standard test conditions (STC): a cloudless day in the mid-seventies with the sun directly overhead. In these idyllic conditions, a panel rated at 200W will produce 200W, but you might not really care because your house will have very low power requirements when you are outside sunbathing. In a less-ideal setting—meaning normally—the panel will probably not produce 200W. But our main concern is to make the panels as efficient as possible and maximize their solar output.

To do that, let's walk outside with a few questions:

- **Where am I?** This is hopefully an easy one, but the answer does dramatically impact your performance. Your typical weather conditions and the amount of sunlight you get over the year determine the light intensity. Diffuse light passing through thin clouds will generate closer to 70W, and thick dark clouds may drag production down to around 20W. And of course, you produce no solar at all during a long winter night.

This is not to discourage the people of the north. Remember, the world's top producer of watts per capita is Germany, a country whose northernmost territories lie at around 54 degrees latitude—parallel with Alaska—and where, in the depths of winter, the sun rises at 8:30 a.m. and sets at 4:00 p.m. That's a mere seven-and-a-half hours of daylight, *maximum*. In fact, virtually all of Germany lies north of Seattle, America's northernmost metropolis. Our major cities sit on latitudes more in line with southern Europe and Northern Africa. Seville, in southwestern Spain, sits about level with Richmond, Virginia. So the length of days and extent of direct sunlight aren't extreme enough anywhere in the country to discourage solar, aside from Alaska.

- **Where can I put them?** Simply taking a look at where to consider installing panels will give you a decent idea of how much direct sunlight they might get. Does the site allow you to face the panels south? Is it shaded by trees, chimneys, or other houses at certain times of day or year?

You also need to keep in mind the proper angle to position the panels. This is an easy decision at the equator, where you simply position the panel horizontally for maximum light. As you move further north, though, you want to keep the panel perpendicular to the sun. You can either mount the devices on a tracking unit, an expensive upgrade that should increase your output by 50 percent in the summer and about 20 percent in the winter, or you can compromise by tilting the panel toward the south (starting from horizontal) at the same angle as your latitude. If you'd rather increase output during the winter versus summer, you could also angle the panel further to the south to even out seasonal performance.

The seasons impact how close you get to your maximum output for another reason. In summer, the sun rises high in the sky, its rays traveling more directly and quickly through less atmosphere. In winter, the lower angle causes sunlight to travel through more of the dust, fog, or pollution that clouds our air pretty much everywhere.

It's not all bad news for northern, colder climes, though. As long as winter snow isn't blocking your panel, the bright, diffuse light reflecting off it supplements the input from the sun's direct rays. At higher temperatures, above the standard operating conditions in which panels are rated, panels actually operate *less* efficiently than at colder ones. In other words, solar photovoltaic technology is successful in Arizona not because it's hotter there, which is an efficiency disadvantage, but because it's practically cloudless.

• **What can I control?** Short of moving your house, you don't really have much choice in the weather or hours of sunlight in your part of the country. In some cases you can increase the amount of daylight your panels receive by cutting off overhanging branches or even removing trees. But the single biggest factor over which you have control is the kind of system and type of photovoltaic panels you install.

Remember: Every silicon cell produces .5 volts, meaning that every panel with thirty-six cells will produce 18 volts. The voltage produced has very little to do with the amount of light radiation and is completely unaffected by opting for a more or less efficient array. But people who pay more for top-of-line panels aren't crazy. The measure we're interested in is watts: voltage per panel multiplied by the current, or amps, per panel.

The classic metaphor for explaining the difference between these two quantities is that voltage is like the water pressure of a hose, while amps are the current or flow of water. You can have all the volts you want, but with no flow you're just holding a hose. So it's the current intensity, or amps, that higher luminosity affects. A panel producing 50 volts (from one hundred cells) times 4 amps will output 200W in prime operating conditions. When your panels receive less-than-optimal light, it's the amp that decreases. The same 50-volt panel producing 3.5 amps is only outputting 175 watts.

Although every manufacturer provides STC wattage ratings, the performance test conditions (PTC) rating may be more useful when sizing up solar panels. In test conditions devised by PV USA in Davis, California, PTC rates solar panel performance in less-than-ideal conditions, that is, the real world. The results are interesting. Some panels rated at 240 STC produce 209 watts in the PTC test, while other more efficient panels produce 226 watts. Doesn't seem like a lot, but think about that 17-watt difference multiplied by, say, five panels, and a difference of 85 watts becomes more significant. At twelve panels, the higher PTC rating becomes equivalent to 226, or almost a whole extra panel. The advantage of the more-efficient panels is that you can, in a sense, buy back some of the wattage lost through diffuse sunlight, shading, available space, and seasonal differences versus other panels. However, the more efficient panels are generally going to cost more than those with lower PTC ratings. Your location, energy needs, roof space, and budget will help determine which panels are right for you.

HYDROPOWER

Compared with solar photovoltaic, water power is refreshingly simple. Flowing water rotates a turbine—basically a glorified medieval mill wheel, except that, instead of processing wheat or corn, the rotation creates electricity via electromagnetic induction.

Electromagnetic induction is also a lot simpler than it sounds. Discovered in 1831 by British scientist Michael Faraday, the process uses the same negative and positive forces that repel and attract magnets to create two magnetic fields, one positive, one negative. The turbine spins something that conducts electricity, like copper wire, between the magnets' opposed fields. The action of the metal spinning between the two fields produces electrons that transfer through the conductive metal to be emitted as electricity. So a turbine is simply copper wire spinning between two magnets. You may even have all these materials at home in a junk drawer or your garage.

Modern power turbines that use electromagnetic induction have been around since Grover Cleveland's first term as president (1885, in case you forgot) and still form the basis of almost all electrical generation today, including nuclear, coal, and natural oil power plants. A residential micro-hydro setup is a time-proven technology, just a cleaner, smaller version of it.

Hydro also has other attractions. Barring drought, water flows all day and all night, producing electricity twenty-four hours a day, unlike solar and wind. If you live in a cooler area with heavy precipitation during the winter, hydro may follow your energy usage cycles better than solar, which has diminishing returns just as you want to crank up the heat in the winter.

Micro-hydro systems are simple and relatively inexpensive. All you need to produce wattage is the turbine and something to channel water into it. However, you have to have the right bit of land, meaning you can figure out your hydro possibilities with one simple question:

- **Do I live near steeply flowing water?** If not, there's only a small chance hydro will work for you. Higher head sites are better because they use smaller pipes and less water. The lower the head, the less efficient and attractive hydropower becomes. Vertical movement of the water is so important that even a powerful but flat river can't easily support a viable hydro system. There is no hydropower to speak of on the Mississippi.

Go out to your water source with a measuring tape, a one-gallon bucket, and a stopwatch to get your numbers before calling a professional. Depending on the figures, it may be a very short conversation—even hydropower advocates admit that it's really practical only for a miniscule number of houses—but at least you'll know what your possibilities and options are.

WIND POWER

Wind, another turbine-based system based on a free natural resource, uses essentially the same technology as water to produce renewable electricity. The wind turns rotors that spin the electrical conductor in a magnetic field. While solar panels can work in almost any unshaded location in the country, wind power relies on a less common natural resource, meaning you can probably sort out your wind power possibilities with a few questions:

- **Which way does the wind blow?** The local wind speed (and, to a lesser extent, direction) is more important than any of the mechanical parts on your turbine. Generally speaking, you'll need regular speeds of between 10 and 25 mph to make a wind system worthwhile. Below 10 mph, the turbine won't produce very much electricity, an amount that decreases and increases exponentially as wind speed goes up.

For example, a site with 12 mph winds might generate 70 percent more energy than a site with a 10 mph breeze.

However, there is an upper limit. Most turbines will shut off above 25 mph to avoid sustaining damage to the equipment, so a generally calm area with powerful but infrequent windstorms won't produce much electricity. Directionally erratic and turbulent winds are also problems.

Residential wind turbines are popping up all over the nation, but you've got the best chance of having adequate winds if you live in the center of the country. The areas with the highest average wind speed cover a swath on the eastern side of the Rockies, stretching from Texas across all of the Great Plains states and most of the Midwest. (Unfortunately, this is also the area of the country most subject to tornadoes.) Generally speaking, the Southeast is the worst area for wind generation, but there are regional variations depending on altitude, topography, and other factors.

You can measure wind speed on your property with an expensive instrument called an anemometer. (Some state energy offices will rent one to you.) Less expensive is visiting Wunderground.com/history and entering your zip code for recent local weather conditions. Select the "Custom" option and enter any date range for your local weather—the past year is a good starting place—and check out your average wind speed, keeping in mind that altitude, hills, and other local topography will make your property's wind speed differ.

Air density is the other environmental element that determines how much wattage you can expect to pull from the sky. The greater the air density, the more power you'll get out of the wind blowing across your rotors, but divining your local air density can be counterintuitive. Humid air, which seems so oppressively heavy, actually has a low density. Air density rises along with barometric pressure and temperatures, while wet conditions are normally associated with lower barometric pressure. Air density also decreases with altitude.

• **What will the neighbors think?** Wind turbines are probably the most aesthetically intrusive renewable energy. To avoid turbulence and maximize speed, the American Wind Energy Association suggests that the bottom edge of the rotor blade reach no lower than thirty feet above the expected height of any trees or buildings within a 500-foot radius. If you've got that kind of space, you're probably not packed into a subdivision or townhouse, but if you're putting ten-foot rotor blades on top of that thirty feet, a forty-foot wind tower sitting unobstructed in the middle of a field is going to be pretty easy to see from a distance. In addition to being highly visible, wind systems can be noisy, and some state and local governments have restrictions on their use.

• **What else do I need?** If you've cleared the natural hurdles, then you can get down to the last two issues that determine your wattage. Like wind speed, a larger rotor blade will generate exponentially more energy.

Think about a pizza. A sixteen-inch pizza costs less and sounds only four inches smaller than a twenty-inch pie, but those four inches are diameter, and the larger size means 60 percent more surface area for olives and pepperoni. Likewise, a ten-foot blade is just two feet longer than an eight-foot blade, but its "swept area" is 60 percent larger, boosting energy production by about 60 percent.

Most turbines are rated in kilowatts, a number derived from a laboratory test. As with solar, power output is frequently substantially different under real world operations, and there is no widely used alternative rating system like PTC for solar panels. In perfect conditions, that 10-kilowatt turbine will produce as much electricity over the course of a year as the average American household uses annually, but wind isn't perfect and is often less predictable than the flow of most water sources.

If you're planning on replacing most of your household power needs with wind, arm yourself with *lots* of local weather data before you decide to invest in a wind turbine.

GEOTHERMAL

Utilizing heat stored deep within the earth is also a centuries-old technology, dating back to thermal baths in which Chinese nobility warmed their third-century BC feet. This hot energy pushes up from the earth's 7,000°C core in the form of magma, which is molten rock that cools slightly as it courses toward the earth's surface but still heats underground water to two or three times its boiling point. This superheated water sometimes escapes as steam at the meeting place between tectonic plates.

Today this steam drives turbines, producing clean, renewable electricity. So geothermal works great as a power source if you live in island nations like Iceland or New Zealand, which sit directly on the earth's deep rifts, marked by volcanoes and earthquakes. But the benefits of residential geothermal aren't all that geographically limited, bringing up the first question you might ask about geothermal heating:

• **What if I don't live near a fault line?** As the residents of California and most of the West Coast will tell you, that's good. The common residential application of geothermal, called binary-cycle, isn't just a scaled-down version of turbine-based power generation and doesn't require tectonic instability in your backyard.

Even at relatively shallow depths, the earth's temperature fluctuates much less than air on the surface. At about thirty feet down, the earth's temperature remains fairly stable at about 50 to 54 degrees, whether it's frigid or baking on the surface. Which generally makes ground temperatures warmer than the air outside during the winter and cooler than surface temperatures in the summer—exactly how we want our houses to feel much of the time.

Residential geothermal can be used by anyone willing to run pipes four or more feet underground, either vertically or horizontally depending on local terrain. If your home sits directly atop a huge stone ridge, sinking pipes straight down is going to be a lot more expensive.

• **What's the catch?** Geothermal doesn't rely on constant wind speeds, clear skies, good southern exposure, or steep watery terrain. You really only need property in which you can dig holes, but the systems do have drawbacks.

Binary-cycle systems pump water or a nontoxic antifreeze liquid from above ground through a loop of pipes laid underground. In the winter, the liquid returns to the surface warmed, and its heat spreads throughout the house in several different ways. A common method uses a heat exchanger to extract the warmth from the liquid. The heat exchanger concentrates the energy, releasing it inside the home via a regular forced air system. It's beautiful technology really, but it's often not worth the money to design a big enough system to provide all your heating needs.

Because of this, geothermal heating often is paired with conventional heating techniques in what's called a dual-source heat pump. These dual-source systems aren't as efficient as full geothermal units, but they cost less to install and have much higher efficiency ratings than conventional furnaces and other heating systems.

Another method for spreading the earth's heat inside a house is called radiant floor—basically what the Romans used two millenniums ago to heat their baths. Heated liquid runs through pipes under your feet, radiating heat up into the room. The warmth spreads evenly across a room and isn't as subject to drafts as forced-air heating. But you have to embed heat pipes in a concrete slab floor—no simple matter.

Yet another system runs geothermally heated water through pipes in the baseboard or radiators, but the heat exchanger needs to bring this water to very high temperatures to heat a room effectively. Also, if you don't have baseboard heating or radiators already, you've got to install a whole new delivery system for your heat. Then again, geothermal systems can also supply a house with hot water.

The real beauty of a binary-cycle pump is that you can reverse the process in the summer. The heat exchanger pulls warm air from the house and transfers it to the circulating liquid. The liquid cools in the earth and pumps back to the surface, where the heat exchanger warms the cooled liquid with warmth from the house and then pumps out the heated liquid, sending it back down to be cooled all over.

The technology sounds exotic, but it's similar to what's keeping your refrigerator cold. A fridge's heat pump pushes out hot air and pulls in new air to be cooled. Because earth-chilled water is generally much cooler than the kitchen-temperature air that fridges suck in, the geothermal system is much more efficient. Although heat exchangers and pumps do use electricity and frequently work in tandem with more conventional heating and cooling systems, they still use much less electricity while controlling your house's temperature. Depending on local geology, the ground loop of pipes can be expensive to install, but it can also last for over fifty years and frequently carries a warranty for much longer than any other piece of renewable energy equipment.

SOLAR THERMAL

The interior of the earth is unimaginably hot, but, unless you're sitting in hot springs, the sun is still the number-one heat source for those of us on the planet's surface. Thus the very simple technology behind solar thermal.

Water or an antifreeze solution feeds into some sort of transparent collector, the sun heats it, and then it runs to where the liquid's energy is used to raise the temperature inside the house, heat showers, or is stored in tanks for later use. Sound familiar? Solar thermal heat can be distributed through the house using essentially the same methods as geothermal systems, including forced air, baseboards, and radiators.

The biggest difference between geothermal and solar thermal is the collector boxes. Flat-plate collectors are the most common and about as simple as they sound. They're typically rectangular boxes with a glass face that allows sunlight to heat the circulating liquid, while maintaining a low profile on a roof.

Another more sophisticated technology is the evacuated-tube collector. where a clear, liquid-filled tube sits within another clear cylinder. All the air between the two tubes is vacuumed out, a process that sounds complicated when compared with pouring circulating liquid in a glass box on the roof. But the vacuuming, or evacuating, can make sense. In a flat-plate collector, the sun heats the liquid, but it also heats the air inside. The circulating air loses heat every time it comes in contact with the cooler glass face. As it continues to circulate, the air conducts the cooler glass temperature to the liquid, cooling it slightly. In an evacuated system, there's no conducting air between the heated liquid and the outer glass tube. The inner tube gets all of the sun's heat with practically no loss.

Of course, it's not cheap to convince air to leave a tube, and evacuated tube systems are more expensive. As with solar photovoltaic, you'll need to make your decision based on energy needs, roof space, location, and budget. Also, most solar thermal systems are not meant to provide more than 80 percent of your heating needs. As with geothermal, designing 100 percent solutions based on solar thermal often isn't practical or cost effective. Building codes may require you to have a back-up heating system at any rate anyway. But using a conventional secondary source for 20 or 30 percent of your heat will have a huge impact on your heating costs.

RESOURCES

In addition to an impressive amount of literature on efficiency and renewable energy sources, there's an enormous amount of information available on the Internet. As with any commodity, lots of websites are trying to sell you on a certain technology or product, or they may have outdated or incorrect information. A solid place to begin your own research on efficiency and renewables for your specific location is the US Department of Energy website:

Energy.gov/basics
Energy.gov/savings

GLOSSARY OF TERMS

Below is a glossary of terms commonly used in discussing energy efficiency, conservation, and generation. (This list has been adapted primarily from the US Department of Energy, Office of Energy Efficiency and Renewable Energy. See www1.eere.energy.gov/site_administration/glossary.html for a more complete list.)

active solar heater
Solar water or space-heating system that uses pumps or fans to circulate water or heat-transfer fluid from solar collectors to a storage tank subsystem.

air retarder/barrier
Material or structural element that inhibits air flow into and out of a building's envelope or shell. A continuous sheet of polyethylene, polypropylene, or extruded polystyrene is wrapped around the outside of a house during construction to reduce the movement of air yet allow water to diffuse easily through it.

air-source heat pump
Heat pump that transfers heat from outdoor air to indoor air during winter and works in reverse during the summer cooling season.

alternating current
Type of electrical current, the direction of which reverses at regular intervals or cycles; in the US the standard is 120 reversals or 60 cycles per second; typically abbreviated as AC (not to be confused with air conditioning, which usually is abbreviated A/C).

alternative fuels
Term for transportation fuels derived from natural gas (propane, compressed natural gas, methanol, etc.) or biomass materials (ethanol, methanol).

anemometer
Instrument for measuring the force or velocity of wind; a wind gauge.

Glossary of Terms

angle of incidence
In reference to solar-energy systems, the angle at which direct sunlight strikes a surface. Sunlight with an incident angle of 90 degrees tends to be absorbed, while lower angles tend to be reflected.

angle of inclination
In reference to solar-energy systems, the angle that a solar collector is positioned above horizontal.

anode
The positive pole or electrode of an electrolytic cell, vacuum tube, etc.

antifreeze
Fluid, such as methanol or ethylene glycol, added to vehicle engine coolant, or used in solar heating system heat transfer fluids, to protect the systems from freezing.

aperture
An opening. In solar collectors, the area through which solar heat is admitted and directed to the absorber.

Appliance Energy Efficiency Ratings
Ratings under which specified appliances convert energy sources into useful energy, as determined by the US Department of Energy.

average demand
The demand on an electrical system or any of its parts over an interval of time as determined by the total number of kilowatt-hours divided by the units of time in the interval.

barrel (petroleum)
42 gallons (306 pounds of oil, or 5.78 million BTUs).

battery
An energy storage device composed of one or more electrolyte cells.

bifacial solar panel
Type of solar panel that uses monocrystalline solar cells as well as thin-film technology on the back so that it captures light on both sides of the panel, boosting efficiency.

Glossary of Terms

bioconversion
Conversion of one form of energy into another by the action of plants or microorganisms, such as the conversion of biomass to ethanol, methanol, or methane.

biomass
As defined by the Energy Security Act of 1980, "Any organic matter which is available on a renewable basis, including agricultural crops and agricultural wastes and residues, wood and wood wastes and residues, animal wastes, municipal wastes, and aquatic plants."

biomass fuel
Biomass converted directly to energy or converted to liquid or gaseous fuels such as ethanol, methanol, methane, and hydrogen.

biomass gasification
The conversion of biomass into a gas.

blown-in insulation (see also loose fill)
Insulation product composed of loose fibers or fiber pellets blown into building cavities or attics using special pneumatic equipment.

British Thermal Unit (BTU)
The amount of heat required to raise the temperature of one pound of water by one degree Fahrenheit; equal to 252 calories.

building orientation
The relationship of a building to true south, as specified by the direction of its longest axis.

capability
The maximum load that a generating unit, power plant, or other electrical apparatus can carry under specified conditions for a given period of time without exceeding its approved limits of temperature and stress.

capacity (heating, of a material)
Amount of heat energy needed to raise the temperature of a given mass of a substance by 1°C. The heat required to raise the temperature of 1 kg of water by 1°C is 4,186 joules.

Glossary of Terms

carbon capture
Also called carbon sequestration, technology designed to prevent large quantities of carbon dioxide produced by fossil-fuel-burning power plants from releasing into the atmosphere by storing them deep in the earth or in the ocean.

carbon dioxide
Colorless, odorless, noncombustible gas with the formula CO_2 present in the atmosphere. The combustion of carbon and carbon compounds (such as fossil fuels and biomass), respiration—a slow combustion in animals and plants—and the gradual oxidation of organic matter in the soil form it.

carbon monoxide
Colorless, odorless, poisonous, combustible gas with the formula CO. The incomplete combustion of carbon and carbon compounds, such as fossil fuels (coal, petroleum) and their byproducts (liquefied petroleum gas, gasoline), and biomass produce it.

catalytic converter
Air-pollution control device that removes organic contaminants by oxidizing them into carbon dioxide and water through a chemical reaction; required in all automobiles sold in America and used in some types of heating appliances.

cathode
Negative pole or electrode of an electrolytic cell, vacuum tube, etc.; the opposite of an anode.

central power plant
Large power plant that generates power for distribution to multiple customers.

chlorofluorocarbons (CFC)
Family of chemicals composed primarily of carbon, hydrogen, chlorine, and fluorine, the principal applications of which are as refrigerants and industrial cleansers. Their principal drawback is a tendency to destroy the earth's protective ozone layer.

circuit
Device or system of devices that allows electrical current to flow through it and allows voltage to occur across positive and negative terminals.

Glossary of Terms

climate change
Term used to describe short- and long-term effects on the earth's climate as a result of human activities, such as fossil-fuel combustion and vegetation clearing and burning.

closed-loop geothermal heat pump system
Closed-loop (also called "indirect") systems circulate a solution of water and antifreeze through a series of sealed loops of piping. Once the heat transfers into or out of the solution, the solution recirculates. Loops can be installed in the ground horizontally or vertically, or they can be placed in a body of water, such as a pond.

collector
Part of a solar-energy heating system that collects solar heat and contains components to absorb that heat and transfer it to a heat-transfer medium (air or liquid).

combustion power plant
A power plant that generates power by combusting a fuel.

compact fluorescent (CFL)
Smaller version of standard fluorescent lamps, in the shape of a light bulb, that directly replaces a standard incandescent light. It consists of a gas-filled tube and a magnetic or electronic ballast.

concentrating (solar) collector
Collector that uses reflective surfaces to concentrate sunlight onto a small area, where it is absorbed and converted to heat or, in the case of solar photovoltaic (PV) devices, into electricity. Concentrators can increase the power flux of sunlight hundreds of times.

conventional fuel
Fossil fuels: coal, oil, and natural gas.

conversion efficiency
Amount of energy produced as a percentage of the amount of energy consumed.

current (electrical)
Flow of electrical energy (electricity) in a conductor, measured in amperes.

demand
Rate at which electricity is delivered to or by a system, part of a system, or piece of equipment expressed in kilowatts, kilovolt-amperes, or other unit at a given instant or averaged over a period of time.

demand charge
Charge for the maximum rate at which energy is used during peak hours of a billing period. The part of a power provider service charged for on the basis of the possible demand as distinguished from the energy actually consumed.

diffuse solar radiation
Sunlight scattered by atmospheric particles and gases so that it arrives at the earth's surface from all directions and cannot be focused.

digester
Device for optimizing the anaerobic digestion of biomass and/or animal manure by bacteria and possibly to recover biogas—primarily composed of methane, carbon dioxide, and hydrogen sulfide—for energy production. Digester types include batch, complete mix, continuous flow (horizontal or plug-flow, multiple-tank, and vertical tank), and covered lagoon.

diode
Electronic device that allows current to flow in one direction only.

direct current
Type of electricity transmission and distribution by which electricity flows in one direction through the conductor; typically abbreviated as DC.

direct normal irradiance (DNI)
Amount of solar heat received at a point on a surface perpendicular (normal) to a straight line to the sun. Solar trackers are sometimes used to maximize the irradiance received by solar photovoltaic or solar thermal installations.

direct solar water heater
These systems use water as the fluid that circulates through the collector to the storage tank. Also known as "open-loop" systems.

double-pane or glazed window
Type of window having two layers (panes or glazing) of glass separated by an

air space. Each layer of glass and surrounding air space reradiates and traps some of the heat that passes through thereby increasing the windows resistance to heat loss. (See **R-value**).

dynamo
Machine for converting mechanical energy into electrical energy by magneto-electric induction; may be used as a motor.

effective capacity
Maximum load that a device can carry.

electric energy
Amount of work accomplished by electrical power, usually measured in kilowatt-hours (kWh). 1 kWh = 1,000 watts = 3,413 BTUs.

electric furnace
Air heater in which air heats over electric resistance heating coils.

electrical system energy losses
Amount of energy lost during the generation, transmission, and distribution of electricity.

electricity
The flow of electrically charged particles, typically electrons, through a conductor.

electricity generation
Producing electricity by transforming other forms or sources of energy into electrical energy; measured in kilowatt-hours.

electrochemical cell
Device containing two conducting electrodes, one positive and the other negative, made of dissimilar materials (usually metals) immersed in a chemical solution (electrolyte) that transmits positive ions from the negative to the positive electrode and thus forms an electrical charge. One or more cells constitute a battery.

electromagnetic energy
Energy generated from an electromagnetic field produced by an electric current flowing through a superconducting wire kept at a low temperature.

electromagnetic field (EMF)
Field created by the presence or flow of electricity in an electrical conductor or electricity consuming appliance or motor.

electron
Elementary particle of an atom with a negative electrical charge and a mass of 1/1837 of a proton; electrons surround the positively charged nucleus of an atom and determine the chemical properties of an atom.

end use
Purpose for which energy or work is consumed.

energy
The capability of doing work. Different forms of energy can be converted to other forms, but the total amount of energy remains the same.

energy audit
Survey that shows the amount of energy in a given place. A helpful exercise to help you find ways to use less energy.

energy charge
The part of an electricity bill based on the amount of electrical energy consumed or supplied.

energy crops
Plants grown specifically for their fuel value. These include food crops, such as corn and sugarcane, and nonfood crops, such as poplar trees and switchgrass.

energy storage
Process of storing, or converting energy from one form to another, for later use; storage devices and systems include batteries, conventional and pumped-storage hydroelectric, flywheels, compressed gas, and thermal mass.

fiberglass
Type of insulation, composed of small-diameter pink, yellow, or white glass fibers, formed into blankets or batts or used in loose-fill and blown-in applications.

First Law of Thermodynamics
Energy cannot be created or destroyed but only changed from one form to another. First Law efficiency measures the fraction of energy supplied

to a device or process that it delivers in its output. Also called the Law of Conservation of Energy.

fissile fuels
Fuel produced by nuclear reactions.

flashing
Typically galvanized sheet metal used to protect against infiltration of precipitation into a roof or exterior wall; usually placed around roof penetrations such as chimneys.

fluorescent light
Conversion of electric power to visible light by using an electric charge to excite gaseous atoms in a glass tube. These atoms emit ultraviolet radiation absorbed by a phosphor coating on the walls of the lamp tube. The phosphor coating produces visible light.

foam (insulation)
A high R-value insulation product usually made from urethane that can be injected into wall cavities or sprayed onto roofs or floors, where it expands and sets quickly.

forced-air system or furnace
Type of heating system in which a fan blows heated air through air channels or ducts to rooms.

fossil fuels
Fuels formed in the ground from the remains of dead plants and animals. It takes millions of years to form fossil fuels, which include oil, natural gas, and coal.

fuel
Any material burned to make energy.

fuel cell
Electrochemical device that converts chemical energy directly into electricity.

fuel efficiency
Ratio of heat produced by a fuel for doing work to the available heat in the fuel.

fuel oil
Any liquid petroleum product burned for the generation of heat in a furnace or firebox or for the generation of power in an engine. Domestic (residential) heating fuels are classed as Nos. 1, 2, and 3; industrial fuels as Nos. 4, 5, and 6.

gasification
Process in which a solid fuel converts into a gas; also known as pyrolitic distillation or pyrolysis.

Gasohol
Registered trademark of a Nebraska state agency for an automotive fuel containing a blend of 10 percent ethanol and 90 percent gasoline.

gasoline
Refined petroleum product suitable for use as a fuel in internal combustion engines.

gas turbine
Turbine in which combusted, pressurized gas flows against a series of blades connected to a shaft, which forces the shaft to turn to produce mechanical energy.

generator
Device for converting mechanical energy to electrical energy.

geothermal energy
Energy produced by the internal heat of the earth. Geothermal heat sources include hydrothermal convective systems, pressurized water reservoirs, hot dry rocks, manual gradients, and magma. Geothermal energy can be used directly for heating or to produce electric power.

geothermal heat pump
Pump that uses the ground, ground water, or ponds as a heat source and heat sink, rather than outside air. Ground or water temperatures are more constant, warmer in winter, and cooler in summer than air temperatures. Geothermal heat pumps operate more efficiently than conventional or air-source heat pumps.

global warming
The increase in average global temperatures due to the Greenhouse Effect.

green certificates, or tags
See renewable energy certificates.

green power
Popular term for energy produced from clean, renewable-energy resources.

green pricing
A practice engaged in by some regulated utilities where electricity produced from clean, renewable resources is sold at a higher cost than that produced from fossil or nuclear power plants because buyers are willing to pay a premium for clean power.

Greenhouse Effect
The heating effect—due to the trapping of long wave (length) radiation—of greenhouse gases produced from natural and human sources.

greenhouse gases
Water vapor, carbon dioxide, tropospheric ozone, methane, and low-level ozone that are transparent to solar radiation but opaque to long-wave radiation, and which contribute to the Greenhouse Effect.

grid
Electricity transmission and distribution system.

grid-connected system
Independent power systems connected to an electricity transmission and distribution system (referred to as the electricity grid) such that the systems can draw on the grid's reserve capacity in times of need and feed electricity back into the grid during times of excess production.

ground loop
In geothermal heat pump systems, a series of fluid-filled (usually plastic) pipes buried in relatively shallow ground (4 to 6 feet deep), or placed in a body of water, near a building. The fluid within the pipes transfers heat between the building and the ground or water in order to heat and cool the building.

ground-source heat pump
See **geothermal heat pump**.

head
Unit of pressure for a fluid, commonly used in water pumping and hydropower to express the height that a pump must lift water or the distance that water falls.

heat
Thermal energy resulting from combustion, chemical reaction, friction, or movement of electricity.

heat-absorbing window glass
Glass that contains special tints that cause the window to absorb as much as 45 percent of incoming solar energy to reduce heat gain in an interior space. Part of the absorbed heat will continue to pass through the window by conduction and reradiation.

heat exchanger
Device used to transfer heat from one medium (liquid or gas) to another medium, but in which the two mediums are physically separated.

heat loss
Heat that flows from a building interior through the envelope to the outside environment.

heat pump
Electricity-powered device that extracts available heat from one area (heat source) and transfers it to another (heat sink) either to heat or cool an interior space or to extract heat energy from a fluid.

heating fuels
Any gaseous, liquid, or solid fuel used for indoor space heating.

heating value
Amount of heat produced from the complete combustion of a unit of fuel.

heating, ventilation, and air conditioning system (HVAC)
All the components of the appliance used to condition the interior air of a building.

home energy rating systems (HERS)
National energy rating program that gives builders, mortgage lenders, secondary lending markets, homeowners, sellers, and buyers a precise evaluation of energy-losing deficiencies in homes. Builders can use this system to gauge the energy quality in their home and also to have a star rating on their home to compare with other similarly built homes.

horizontal ground loop
In this type of closed-loop geothermal heat pump installation, the fluid-filled plastic heat exchanger pipes are laid out in a plane parallel to the ground surface. The most common layouts either use two pipes, one buried at six feet, and the other at four feet, or two pipes placed side-by-side at five feet in the ground in a two-foot-wide trench. Horizontal ground loops generally are most cost-effective for residential installations, particularly for new construction where sufficient land is available.

hybrid system
Renewable energy system that includes two different types of technologies that produce the same type of energy; for example: a wind turbine and a solar photovoltaic array combined to meet a power demand.

hydroelectric power plant
Power plant that produces electricity by the force of water falling through a hydro turbine that spins a generator.

hydrogen
Colorless, odorless, highly combustible gas with the formula H_2 present in the atmosphere. Its high-energy content means that, in liquid form, it can be used as a fuel.

ignition point
Minimum temperature at which the combustion of a material, solid or fluid, can occur.

incandescent
These lights use an electrically heated filament to produce light in a vacuum or inert gas-filled bulb.

indirect solar gain system
Passive solar heating system in which the sun warms a heat storage element

and the heat is distributed to the interior space by convection, conduction, and radiation.

indirect solar water heater
These systems circulate fluids other than water (such as diluted antifreeze) through the collector. The collected heat transfers to the household water supply using a heat exchanger. Also known as "closed-loop" systems.

infrared radiation
Electromagnetic radiation (heat) capable of producing a thermal or photovoltaic effect, though less effective than visible light.

insolation
The solar power density incident on a surface of stated area and orientation, usually expressed as watts per square meter or BTUs per square foot per hour. Not to be confused with insulation.

installed capacity
Total capacity of electrical generation devices in a power station or system.

insulation blanket
Pre-cut layer of insulation applied around a water-heater storage tank to reduce standby heat loss.

inverter
Device that converts direct current electricity (from, for example, a solar photovoltaic module or array) to alternating current to operate appliances directly or to supply power to an electricity grid.

investor owned utility (IOU)
Power provider owned by stockholders or other investors; sometimes referred to as a private power provider, in contrast to a public power provider owned by a government agency or cooperative.

ion
Electrically charged atom or group of atoms that has lost or gained electrons; a loss makes the resulting particle positively charged; a gain makes the particle negatively charged.

irradiance
The direct, diffuse, and reflected solar radiation that strikes a surface.

kilowatt (kW)
Standard unit of electrical power equal to 1,000 watts or the energy consumption of 1,000 joules per second. 1,000 kilowatts = 1 megawatt. 1 million kilowatts = 1 gigawatt. 1 billion kilowatts = 1 terawatt.

kilowatt-hour (kWh)
Measure of electricity supply or consumption of 1,000 watts over the period of one hour; equivalent to 3,412 BTUs.

leaking electricity
Related to standby power, leaking electricity is the power needed for electrical equipment to remain ready for use while in a dormant mode or operation. Many electrical devices, such as TVs, stereos, and computers are consuming this kind of electricity even when you think they are off. See **phantom load**.

lithium-sulfur battery
Battery that uses lithium in the negative electrode and a metal sulfide in the positive electrode, with molten salt as the electrolyte; it can store large amounts of energy per unit weight.

load
Power required to run a defined circuit or system, such as a refrigerator, building, or an entire electricity distribution system.

load factor
Ratio of average energy demand (load) to maximum demand (peak load) during a specific period.

load leveling
Deferment of certain loads to limit electrical power demand, or the production of energy during off-peak periods for storage and use during peak demand periods.

loose-fill
Insulation made from rock wool, fiberglass, cellulose, vermiculite, or perlite;

composed of loose fibers or granules applied by pouring directly from the bag or with a blower.

low-emissivity coatings and (window) films
Coating applied to the surface of the glazing of a window to reduce heat transfer.

low-emissivity windows and (window) films
Energy-efficient windows with the coating or film above applied to the surface of the glass.

low-flush toilet
Toilet that uses less water than a standard one during flushing for the purpose of conserving water.

lumens per watt (LPW)
Measure of the efficiency of lamps that indicates the amount of light (lumens) emitted by the lamp for each unit of electrical power used.

magma
Molten or partially molten rock. Some magma bodies are believed to exist at drillable depths within the earth's crust, although practical technologies for harnessing magma energy have not been developed. If ever utilized, magma represents a potentially enormous resource.

mean wind speed
The average wind speed over a specified time period and height above the ground. The majority of National Weather Service anemometers are 20 feet tall.

methane
Colorless, highly flammable gas with the formula CH_4. The main constituent of natural gas, it can be formed naturally by bacteria or can be manufactured.

micro-hydropower
Water power adapted for small-scale use, such as electricity for a residence. These systems usually generate up to 100 kW of electricity. See also **hydroelectric power plant**.

monocrystalline
Type of solar cell produced from single-crystal wafer cells. They tend to be more expensive but more efficient than other crystalline cells.

municipal solid waste
Waste material from households and businesses in a community that is not regulated as hazardous.

municipal waste to energy project (or plant)
Facility that produces fuel or energy from municipal solid waste.

National Electrical Code (NEC)
Set of regulations that have contributed to making the electrical system in America one of the safest in the world. The National Fire Protection Association has sponsored the NEC since 1911, and the NEC is updated every three years.

natural gas
Hydrocarbon gas obtained from underground sources, often in association with petroleum and coal deposits. It generally contains a high percentage of methane, varying amounts of ethane, and inert gases; used as a heating fuel.

nonrenewable fuels
Fuels that cannot be easily made or "renewed," such as oil, natural gas, and coal.

normal recovery capacity
Characteristic of domestic water heaters that is the amount of gallons raised 100°F per hour (or minute) under a specified thermal efficiency.

nuclear energy
Energy derived from splitting atoms of radioactive materials, such as uranium, and which produces radioactive waste in the form of spent-fuel rods.

ocean energy systems
Conversion technologies that harness the energy in tides, waves, and thermal gradients in the oceans.

ocean thermal energy conversion (OTEC)
The process of or technologies for producing energy by harnessing the temperature differences (thermal gradients) between surface waters and ocean depths.

off-peak
Period of low energy demand, as opposed to maximum, or peak, demand.

ohms
Measure of the electrical resistance of a material equal to the resistance of a circuit in which the potential difference of 1 volt produces a current of 1 ampere.

oil (fuel)
Product of crude oil used for space heating, diesel engines, and electricity generation.

open-loop geothermal heat pump system
Open-loop (also known as "direct") systems circulate water drawn from a ground or surface water source. Once the heat transfers into or out of the water, the water returns to a well or surface discharge instead of recirculating through the system. This option is practical where there is an adequate supply of relatively clean water, and all local codes and regulations regarding groundwater discharge are met.

orientation
Alignment of a building along a given axis to face a specific geographical direction. The alignment of a solar collector in degrees east or west of true south.

oxygenates
Gasoline fuel additives such as ethanol, ETBE, or MTBE that add extra oxygen to gasoline to reduce carbon monoxide pollution produced by vehicles.

panel (solar)
Individual solar photovoltaic collectors or modules.

parabolic dish
Solar energy-conversion device with a bowl-shaped dish covered with a highly reflective surface that tracks the sun and concentrates sunlight on a

fixed absorber, thereby achieving high temperatures, for process heating or to operate a heat engine to produce power or electricity.

parabolic trough
Solar energy conversion device that uses a trough covered with a highly reflective surface to focus sunlight onto a linear absorber containing a working fluid that can be used for medium temperature space or process heat or to operate a steam turbine for power or electricity generation.

particulates
Fine liquid or solid particles in combustion gases. The Environmental Protection Agency regulates the quantity and size of particulates emitted by cars, power and industrial plants, wood stoves, etc.

passive solar (building) design
Architectural design that uses structural elements to heat and cool a building without the use of mechanical equipment.

passive solar heater
Solar water or space-heating system in which solar energy is collected and/or moved by natural convection without pumps or fans.

payback period
Amount of time required before the savings resulting from your system equal total system cost.

peak demand (or load)
Maximum energy demand or load in a specified time period.

peak energy
The rated amount of power that solar PV (photovoltaic) panels will produce under ideal conditions.

peak sun hours
The equivalent number of hours per day when solar irradiance averages 1 kW/m^2 — that is, one kilowatt per square meter on which the sun falls. For example, six peak sun hours mean that the energy received during total daylight hours equals the energy that would have been received had the irradiance for six hours been 1 kW/m^2.

peaking capacity
Power generation equipment or system capacity to meet peak power demands.

pellet stove
Space-heating device that burns pellets; they're more efficient, clean burning, and easier to operate relative to conventional wood-burning appliances.

pellets
Solid fuel made primarily from wood sawdust compacted under high pressure to form small pellets (about the size of rabbit feed) for use in a pellet stove.

performance ratings
Solar collector thermal performance ratings based on collector efficiencies, usually expressed in BTUs per hour.

phantom load
The power that any appliance consumes when turned off but not unplugged. Examples of phantom loads include appliances with electronic clocks or timers, with remote controls, and with wall cubes (a small box that plugs into an AC outlet to power appliances).

phase-change material
Material that can be used to store thermal energy as latent heat. Various types of materials have been and are being investigated—such as inorganic salts, eutectic compounds, and paraffins—for a variety of applications, including solar-energy storage. (Solar energy heats and melts the material during the day, and at night it releases the stored heat and reverts to a solid state.)

photovoltaic (solar) module or panel
Product that generally consists of groups of PV cells electrically connected together to produce a specified power output under standard test conditions, mounted on substrate, sealed with encapsulant, and covered with protective glazing. It may be further mounted on an aluminum frame. A junction box on the back or underside of the module allows for connecting the module circuit conductors to external conductors.

pitch control
Method of controlling a wind turbine's speed by varying the orientation, or pitch, of the blades, thereby altering its aerodynamics and efficiency.

power
Energy capable or available for doing work; the time rate at which work is performed, measured in horsepower, watts, or BTUs per hour. Electric power is the product of electric current and electromotive force.

power coefficient
Ratio of power produced by a wind-energy conversion device to the power in a reference area of the free windstream.

power generation mix
Relative proportion of electricity distributed by a power provider that is generated from available sources such as coal, natural gas, petroleum, nuclear, hydropower, wind, or geothermal.

power purchase agreements (PPA)
Contracts in which a power utility agrees to provide energy from renewable sources (such as solar or wind) to a customer, normally providing a fixed rate for the power purchased over the length of the contact.

power (solar) tower
Solar thermal central receiver power systems, where an array of reflectors focuses sunlight onto a central receiver and absorber mounted on a tower.

programmable thermostat
Electronic control box that allows you to program into the device's memory a preset schedule of times (when certain temperatures occur) to turn on HVAC equipment.

propane
A hydrocarbon gas, C_3H_8, occurring in crude oil, natural gas, and refinery cracking gas. It is used as a fuel, solvent, and refrigerant. Propane liquefies under pressure and is the major component of liquefied petroleum gas (LPG).

propeller (hydro) turbine
Turbine that has a runner with attached blades, similar to a propeller used to drive a ship. As water passes over the curved blades, it causes the shaft to rotate.

PV
Abbreviation of photovoltaic.

radiant floor
Heating system in which the floor contains channels or tubes through which hot air or water circulates and the whole floor heats evenly. Thus, the room heats from the bottom up. Radiant floor heating eliminates the draft and dust problems associated with forced air heating systems.

radiant heating system
Means by which heated surfaces—such as electric resistance elements, hot water (hydronic) radiators, etc.— supply (radiate) heat to a room.

radiation
Transfer of heat through matter or space by means of electromagnetic waves.

radioactive waste
Materials left over from making nuclear energy. Radioactive waste can harm living organisms if not stored safely.

rated life
Length of time that a product or appliance is expected to meet a certain level of performance under normal operating conditions; in a light fixture or fitting, the period after which the lumen depreciation (lamp failure) is at 70 percent of its initial value.

rate schedule
Mechanism used by utilities to determine prices for electricity; typically defines rates according to amounts of power demanded/consumed during specific time periods.

recycling
Converting materials no longer useful as designed or intended into a new product.

reflective coatings
Materials applied to window glass before installation. These coatings reduce radiant heat transfer through windows and also reflect outside heat and a portion of the incoming solar energy, thus reducing heat gain.

reflective window films
Material applied to window panes that controls heat gain and loss, reduces glare, minimizes fabric fading, and provides privacy. These films are retrofitted on existing windows.

refrigerant
Fluid used in air conditioners, heat pumps, and refrigerators to transfer heat into or out of an interior space. This fluid boils at a very low temperature enabling it to evaporate and absorb heat easily.

renewable energy
Energy derived from regenerative resources that for practical purposes cannot be depleted. Types of renewable energy resources include moving water (hydro, tidal, and wave power), thermal gradients in ocean water, biomass, geothermal energy, solar energy, and wind energy. Municipal solid waste is also considered to be a renewable energy resource.

renewable energy certificates (RECs).
Also known as green tags or tradeable renewable certificates, RECs represent the environmental attributes of power produced from renewable resources. By separating the environmental attributes from the power, clean-power generators are able to sell the electricity they produce to power providers at competitive market values. The additional revenue generated by the sale of the green certificates covers the above-market costs associated with producing power made from renewable energy sources.

resistance
Inherent characteristic of a material to inhibit the transfer of energy. In electrical conductors, electrical resistance results in the generation of heat. Electrical resistance is measured in ohms. The heat transfer resistance properties of insulation products are quantified as the R-value.

R-value
Capacity of a material to resist heat transfer. The R-value is the reciprocal of a material's conductivity (U-value). The higher the R-value of a material, the greater its insulating properties.

seasonal energy efficiency ratio (SEER)
Measure of seasonal or annual efficiency of a central air conditioner or air

conditioning heat pump. It takes into account the variations in temperature that can occur within a season and averages BTUs of cooling delivered for every watt-hour of electricity used by the heat pump over a cooling season.

seasonal performance factor (SPF)
Ratio of useful energy output of a device to the energy input, averaged over an entire heating season.

Second Law of Thermodynamics
No device can completely or continuously transform all of the energy supplied to it into useful energy.

semiconductor
Any material with a limited capacity for conducting electric current. Certain semiconductors, including silicon, gallium arsenide, copper indium diselenide, and cadmium telluride, are uniquely suited to the photovoltaic conversion process.

setback thermostat
Thermostat that can be set to automatically lower temperatures in an unoccupied house and raise them again before the occupant returns.

shading coefficient
Measure of window glazing performance that is the ratio of the total solar heat gain through a specific window to the total solar heat gain through a single sheet of double-strength glass under the same set of conditions; expressed as a number between 0 and 1.

silicon
Semimetallic element and excellent semiconductor material used in solar photovoltaic devices; commonly found in sand. Silicone is a synthesized form.

single glaze or pane
One layer of glass in a window frame. It has very little insulating value (R-1), provides only a thin barrier to the outside, and can account for considerable heat loss and gain.

Slinky™ ground loop
In this type of closed-loop horizontal geothermal heat pump installation, fluid-filled plastic heat exchanger pipes are coiled like a spring to allow more pipe in a shorter trench. This type of installation cuts down on installation costs and makes horizontal installation possible in unconventional areas.

smart window
Technologically advanced window system that contains glazing that can change or switch its optical qualities when a low voltage electrical signal is applied to it, or in response to changes in heat or light.

solar air heater
Solar thermal system in which air heats in a collector and either transfers directly to the interior space or to a storage medium, such as a rock bin.

solar array
Group of solar collectors, modules, or reflectors connected together to provide electrical or thermal energy.

solar collector
Device used to collect, absorb, and transfer solar energy to a working fluid. Flat-plate collectors are the most common type used for solar water or pool heating systems. In a photovoltaic system, the solar collector could be crystalline silicon panels or thin-film roof shingles, for example.

solar energy
Electromagnetic energy transmitted from the sun. The amount that reaches the earth equals one billionth of total solar energy generated, or about 420 trillion kilowatt-hours.

solar film
Window glazing coating, usually tinted bronze or gray, used to reduce building cooling loads, glare, and fabric fading.

solar furnace
Device that achieves very high temperatures by using reflectors to focus and concentrate sunlight onto a small receiver.

solar hot-water heaters
Systems designed to heat water with the sun's energy for delivery around a residence or other building. Solar hot water heaters can either be stand-alone systems or used in conjunction with other types of water heaters.

solar panel
See **photovoltaic (solar) module or panel**.

solar space heater
Solar energy system designed to provide heat to individual rooms in a building.

solar thermal
Literally, any system that converts solar energy into heat. For residential purposes, solar thermal generally refers to passive systems that use flat-plate solar collectors to heat a liquid that is then circulated to raise the temperature inside the house.

solid fuels
Any fuel in solid form, such as wood, peat, lignite, coal, and manufactured fuels such as pulverized coal, coke, charcoal, briquettes, pellets, etc.

standby power
For the consumer, the electricity used by your TVs, stereo, and other electronic devices that use remote controls. When you press "off," minimal power (dormant mode) is still being used to maintain the internal electronics in a ready, quick-response mode. This way, your device can be turned on with your remote control and be immediately ready to operate. See **phantom load**.

steam
Water in vapor form; used as the working fluid in steam turbines and heating systems.

steam boiler
Type of furnace in which fuel is burned and the heat is used to produce steam.

steam turbine
Device that converts high-pressure steam, produced in a boiler, into mechanical energy used to produce electricity by forcing blades in a cylinder to rotate and turn a generator shaft.

super window
Highly insulating window with a heat loss so low it performs better than an insulated wall in winter, since the sunlight that it admits is greater than its heat loss over a 24-hour period.

surface water loop
In this type of closed-loop geothermal heat pump installation, the fluid-filled plastic heat exchanger pipes are coiled into circles and submerged at least 8 feet below the surface of a body of surface water, such as a pond or lake.

tankless water heater
Water heater that heats water immediately before it is directly distributed for end use; also called an on-demand water heater.

temperature coefficient (of a solar photovoltaic cell)
Amount that the voltage, current, and/or power output of a solar cell changes due to a change in the cell temperature.

therm
Unit of heat equivalent to 100,000 BTUs.

thermal efficiency
Measure of the ability to convert a fuel to energy and useful work.

thermal energy
Any energy developed through the use of heat.

thermal envelope houses
An architectural design (also known as the double envelope house), sometimes called a house within a house, that employs a double envelope with a continuous airspace of at least 6 to 12 inches on the north wall, south wall, roof, and floor, achieved by building inner and outer walls, a crawl space or sub-basement below the floor, and a shallow attic space below the weather roof. The east and west walls are single, conventional walls. A buffer zone of solar-heated, circulating air warms the inner envelope of the house. The south-facing airspace may double as a sunspace or greenhouse.

thermal resistance (R-value)
This designates the resistance of a material to heat conduction. The larger the R-value number, the greater the resistance.

thermodynamics
Study of the transformation of energy from one form to another and its practical application. See **First** and **Second Law of Thermodynamics**.

thermostat
Device that controls the operation of heating and cooling devices by turning a heat source on or off when a specified temperature occurs.

thin-film
Layer of semiconductor material, such as copper indium diselenide or gallium arsenide, a few microns in thickness or less, used to make solar photovoltaic cells.

tidal power
Energy available from the rise and fall of ocean tides. A tidal power plant works on the principal of a dam or barrage that captures water in a basin at the peak of a tidal flow, then directs the water through a hydroelectric turbine as the tide ebbs.

tracking solar array
Array that follows the path of the sun to maximize solar heat incident on the PV surface. The two most common orientations are (1) one-axis tracking where the array tracks the sun east to west, and (2) two-axis where the array points directly at the sun at all times. Tracking arrays use both direct and diffuse sunlight. Two-axis tracking arrays capture the maximum possible daily energy.

true south
Direction, at any point on the earth in the northern hemisphere, facing toward the South Pole. Essentially a line extending from the point on the horizon to the highest point that the sun reaches on any day (solar noon) in the sky.

turbine
Device for converting the flow of a medium (hot gases, air, steam, water) into mechanical motion.

unglazed solar collector
Solar thermal collector with an absorber that doesn't have a glazed covering. Solar swimming pool heater systems usually use unglazed collectors because they circulate relatively large volumes of water through the collector and capture nearly 80 percent of available solar energy.

utility
A regulated entity that often exhibits the characteristics of a natural monopoly. For the purposes of electric industry restructuring, "utility" refers to the regulated, vertically integrated electric company. "Transmission utility" refers to the regulated owner/operator of the transmission system only. "Distribution utility" refers to the regulated owner/operator of the distribution system that serves retail customers.

U-value
The reciprocal of R-value. The lower the number, the greater the heat transfer resistance (insulating) characteristics of the material.

vampire load
See **phantom load**.

vertical ground loop
In this type of closed-loop geothermal heat pump installation, the fluid-filled plastic heat exchanger pipes are laid out in a plane perpendicular to the ground surface. Holes approximately 4 inches in diameter are drilled about 20 feet apart and 100 to 400 feet deep. Into these holes go two pipes connected at the bottom with a U-bend to form a loop. The vertical loops connect with horizontal pipe (manifold), are placed in trenches, and connect to the heat pump in the building.

Large commercial buildings and schools often use vertical systems because the land area required for horizontal ground loops would be prohibitive. Vertical loops are also used where the soil is too shallow for trenching, or for existing buildings, since they minimize the disturbance to landscaping.

volt (V)
Unit of electrical force equal to the electromotive force that will cause a steady current of 1 ampere to flow through a resistance of 1 ohm.

voltage
The amount of electromotive force, measured in volts, that exists between two points.

wafer
Thin sheet of (photovoltaic) semiconductor made by cutting it from a single crystal or ingot.

water-source heat pump
Type of geothermal heat pump that uses ground or surface water as a heat source. Water has a more stable seasonal temperature than air, making for a more efficient heat source.

water turbine
Turbine that uses water pressure to rotate its blades; primarily used to power an electric generator.

water wall
Interior wall made of water-filled containers for absorbing and storing solar energy.

watt
Rate of energy transfer equivalent to 1 ampere under an electrical pressure of 1 volt. 1 watt = 1/746 horsepower = 1 joule per second. It is the product of voltage and current (amperage).

wave power
Capturing the motion of ocean waves and converting it into energy.

weatherization
Caulking and weatherstripping to reduce a building's air infiltration and exfiltration.

weatherstripping
Thick, felt-like material used to seal gaps around windows and exterior doors.

wind energy
Energy available from the movement of the wind across a landscape caused by the heating of the atmosphere, earth, and oceans by the sun.

wind-power farm
Group of wind turbines interconnected to a common power provider system through a system of transformers, distribution lines, and (usually) one substation. Operation, control, and maintenance functions are often centralized through a network of computerized monitoring systems, supplemented by visual inspection.

wind speed
Rate of flow of the wind, undisturbed by obstacles.

wind turbine
Wind energy conversion device that produces electricity. Unlike a windmill, its sole purpose is to generate energy, not turn a mill.

wind turbine-rated capacity
Amount of power a wind turbine can produce at its rated wind speed, for example: 100 kW at 20 mph. The rated wind speed generally corresponds to the point at which the conversion efficiency nears its maximum. Because of the variability of the wind, the amount of energy a wind turbine actually produces is a function of the capacity factor (that is, a wind turbine produces 20 to 35 percent of its rated capacity over a year).

wind velocity
Speed and direction of wind in an undisturbed flow.

zone
Area within the interior space of a building, such as an individual rooms, to be cooled, heated, or ventilated.

zoning
Combining rooms in a structure according to similar heating and cooling patterns. Zoning requires using more than one thermostat to control heating, cooling, and ventilation equipment.

ACKNOWLEDGMENTS

Green Is Good has been a family affair literally decades in the making. My brother Tom recruited me for my first political campaign when I was in fourth grade—a successful effort to elect Joe Moakley to Congress. My sister Kelly hooked me into my first environmental campaign in the late 1970s. Her infectious energy and enthusiasm made "The Bottle Bill" seem almost as exciting as the Moakley victory. By the time the Bottle Bill passed in 1981, I was hooked. SmartPower's campaigns today still build on Tom and Kelly's examples.

Tom and another brother, Jim, rode herd over the writing of this book as well. Their, um, forceful criticisms—delivered as only older brothers can—have made this book far better than it would have been otherwise.

And as should be obvious, no one taught us better than my parents, Tom and Betty Ann Keane. Two more loving, grounded and perfect parents you could not meet. They created a family and a childhood for us that is second to none, and today they continue to live their lives as true examples to me and all my brothers and sisters.

Of course, the great minds and funders who created SmartPower back in 2002 really deserve the credit for the contents of this book. Among them: Ruth Hennig of The John Merck Fund, Bryan Garcia of the Clean Energy Finance and Investment Authority (formerly the Connecticut Clean Energy Fund), Stewart Hudson of the Emily Hall Tremaine Foundation, Michael Northrop of The Rockefeller Brothers Fund, Lea Aeschleman of The Pew Charitable Trusts, and

Camilla Seth of the Surdna Foundation. Without their support and vision, SmartPower and *Green Is Good* never would have gotten off the ground. Credit goes as well to the SmartPower Board of Directors: Gary Simon, Douglas Foy, Robin Rather, Tom Rawls, and the aforementioned Ruth Hennig. Their guidance and commitment have been critical not just to this book, but to our cause.

Truth in packaging: The SmartPower staff throughout the years really created these stories. Lyn Rosoff and Jonathan Edwards were there since day one, and together we actually lived through all that I write about. It's no exaggeration to say that the success we've seen at SmartPower is completely due to their dedication and commitment. Over the years we've benefited from the guidance and passion of a host of people, among them: Brian Wright, Toni Bouchard, Daniel Francis, Chandler Clay, Dru Bacon, Molly Tsongas, Keri Enright, Bob Wall, Aparna Mohla, Kate Kelly, Pat Doherty, Marissa Newhall, Andrew Voris, Bernadette Buck, Amy Barber, Greg Fisher, Paul and Kim Pita, Tom Shea, Lise Dondy, Willy Ritch, Elizabeth Reich Keane, Steve Zulli, the geniuses at Genius Pool, and an enthusiastic army of interns and partners who joined together to prove to the American people that clean energy today is real, it's here, and it's working.

A special thanks also to Eran Mahrer, formerly of Arizona Public Service and now with The Solar Electric Power Association (SEPA), for his faith in SmartPower and all we can achieve. He shares my view that we're just getting started.

Acknowledgments

The writing of this book wouldn't have happened without the guidance of my agent, David Patterson at Foundry Media. Dylan Lowe at West Wing Writers made the introduction, and Howard Means guided me each step of the way. Nathan Means and his expertise really added depth and value. James Jayo at Lyons Press brought the entire project together with focus and direction. And Meredith Dias saw the book to the finish line with grace and intelligence.

I also need to thank the two writers I most admire—my in-laws Jon and Kem Sawyer. They are beautiful writers in every way. Thanks for your encouragement and hand holding along the way, and for everything you've given me.

Credit, too, must go to all those mentioned in *Green Is Good*. It's the people in the stories herein who are changing our society for the better. They are the ones proving every day that green is good . . . and getting better.

And finally to my wife, Kate, who lives and breathes SmartPower with me every day. I can't imagine life without you. Green is good—but you're the best.

INDEX

Index

Index

Index

Index

Index

Index

Index

NOTES

NOTES